OPERATOR THEORY, ANALYTIC FUNCTIONS MATRICES, AND ELECTRICAL ENGINEERING

Conference Board of the Mathematical Sciences
REGIONAL CONFERENCE SERIES IN MATHEMATICS

supported by the
National Science Foundation

Number 68

OPERATOR THEORY, ANALYTIC FUNCTIONS MATRICES, AND ELECTRICAL ENGINEERING

J. William Helton
with the assistance of
Joseph A. Ball, Charles R. Johnson
and John N. Palmer

Published for the
Conference Board of the Mathematical Sciences
by the
American Mathematical Society
Providence, Rhode Island

Expository Lectures
from the CBMS Regional Conference
held at Lincoln, Nebraska
August 1985

Research supported in part by National Science Foundation Grant DMS-83-01394 and Office of Naval Research N00014-85-K-0070.

1980 *Mathematics Subject Classifications* (1985 *Revision*). Primary 47-01, 46-01, 93-XX, 22-XX, 30-XX, 42-XX, 49-XX.

Library of Congress Cataloging-in-Publication Data

Helton, J. William, 1944–
    Operator theory, analytic functions, matrices, and electrical engineering.
    (Regional conference series in mathematics, ISSN 0160-7642; no. 68)
    Bibliography: p.
    1. Functional analysis.   2. Operator theory.   3. Matrices.   I. Conference Board of the Mathematical Sciences.   II. Title.   III. Series.
    QA1.R33 no. 68   [QA321]   510 s [515.7]   87-1192
    ISBN 0-8218-0718-8

Contents

Part I. Engineering motivation

What is the transfer function—how analytic functions arise in engineering-connection laws.

Engineering problems (vs. analysis); a control paradigm.

Part II. Analytic function theory

How much classical analysis can you get by staring at pairs of invariant subspaces? There is a unified way of obtaining the theories of classical interpolation, H^∞ approximation, Corona = Bezout identities, commutant lifting, Wiener-Hopf factorization, integrable systems (Toda Lattice, KdV), matrix LU decompositions, interpolation, upper triangular approximation. The method gives excellent results when all functions induced are differentiable and in these cases frequently extends existing results in various ways, e.g., treats added symmetries, allows poles in H^∞ functions.

Basic projective lore. The correspondences between linear fractional actions on operators and linear actions on subspaces.

A general theory which contains both the matrix and analytic function case.

Preface

These notes expand ten lectures given at a regional conference in Lincoln, Nebraska. The objective is to describe the idea behind a broad variety of topics in a brief volume. The conference assembled a wide variety of scientists: pure mathematicians on one hand and engineers on the other. Consequently one of the constraints on these notes was that they stake out a middle ground and make the extremes accessible to most conference participants. However, these extremes provided a distinctive flavor which is missing from these notes. In particular talks by other participants gave substantial practice in applying the results of Chapter 5 and Chapter 8 to control problems.

This volume splits into four parts which are nearly independent. An approximate description of what you must have read in order to read a particular chapter is:

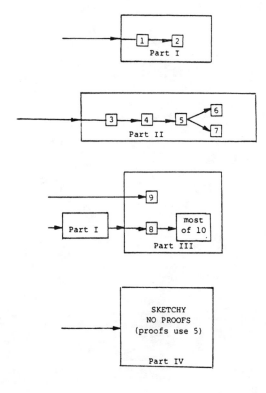

Many readers might want to skip Part I entirely and go directly to Part II.

While the impetus to write these notes was a ten lecture series, they correspond more closely to the 30-lecture graduate topics course in which they were developed. I think that a reasonable topics course could be based on sections of these notes. A reasonable plan is as follows:

• Part I supplemented with engineering texts Dorf [**Dor**] or Ogata [**O**] which are very easy to read.

• In Part II do 3, 4, 5, 6.A, 6.D, 6.E, 6.F, a quick pass at 7, or do [**BG**] in detail. General supplemental references are [**Hls**], [**D**], [**Du1-Du2**], [**G**], [**RR2**], while specialized references are [**S1-S2**], [**BH1–BH5, 8**], [**BG**], [**SW1-SW2**], and [**Y**].

• Parts III and IV as they stand are a bit sketchy for classroom use.

A warning is that throughout these notes the terms chapter and homework exercise are used with tongue in cheek. Usually what we term chapter requires several chapters and the homework would be awfully rough on a beginner.

These notes were written by several people. Joe Ball and I wrote Chapter 6.A and Chapter 8. Charlie Johnson wrote Chapter 9. Chapter 6.F is based mostly on lectures and discussions with John Palmer and somewhat on discussions with Nolan Wallach. Also John Doyle checked Chapter 10 while Bruce Francis checked Chapters 1 and 2. My students Dave Schwartz and Orlando Merino checked most of the manuscript; they made valuable remarks. Neola Crimmins did a marvelous job of typing the original manuscript and implementing many revisions.

Notation Guide

Chapter 4

Chapter 5

Chapter 8

Chapter 9

Chapter 10

$\mu_{\mathbf{Q}}(W)$

The spectral radius of W w.r.t. the set   111
$\mathbf{Q}$ $\sup_{\xi \in \sigma_{\mathbf{Q}}(W)} \inf_x \frac{\|\xi x\|}{\|x\|}$.

Chapter 11

(OPT)

Find $\inf_{f \in A_N} \sup_{\theta} \Gamma(e^{i\theta}, f(e^{i\theta}))$, where   121
$\Gamma(e^{i\theta}, z)$ is a nonnegative-valued function of
$e^{i\theta}$ and $z \in \mathbb{C}^N$.

$\kappa(A)$

The condition number   123
$\sup_{\vec{h}} \{ \inf_{\theta} | \sum_j a_j h_j : \vec{h} \in \overline{\mathbf{BA}}_N$
wno $\sum_j a_j h_j \leq 0 \}$ where $\mathbf{A}_N = $ set of
N-vector valued functions analytic and
continuous on the disk
$\overline{\mathbf{BA}}_N = \{ \vec{h} \in \mathbf{A}_N : \sum_j |h_j|^2 \leq 1 \}$, and
$a_j(e^{i\theta}) = (\partial \Gamma / \partial z)(e^{i\theta}, f_0(e^{i\theta}))$.

Throughout $\mathbf{B}$ is a prefix which stands for open unit ball, e.g., $\mathbf{B}L^2 = \{ j \in L^2 : \|f\|_{L^2} < 1 \}$. For the closed unit ball we use $\overline{\mathbf{B}}$. Also L_N^p, H_N^p sometimes denotes a function space on the circle and sometimes a function space on the imaginary axis. There should be little trouble in telling which usage is intended in any particular context. After Chapter 2 the circle is used almost exclusively.

Dedicated to
Joanne
Maxene and John Helton
Frances and Jim

Part I. Engineering Motivations

1. Engineering Background

The first chapter is designed to convince a mathematician that analytic functions $f(s)$ occur naturally in describing a "black box," that is s does indeed correspond to frequency, and that stable boxes (a very important class) give $f(s)$ with no poles in the right half-plane (R.H.P.). These are modest expectations of a long chapter and fastidious attention to detail is not essential. In fact beyond Chapter 2, the first chapter is not necessary.

Much of engineering concerns boxes which take in inputs and put out outputs. In these talks the inputs will be in the vector space $\mathbb{C}^n$ and the outputs in $\mathbb{C}^m$, and they vary with time. Thus a box is really a map B from a space of $\mathbb{C}^n$-valued functions to $\mathbb{C}^m$-valued functions (see Figure 1.1).

Common physical properties of a box are

1. *Causality*. If two inputs $i_1(t)$ and $i_2(t)$ are identical before time t_0, then the outputs are identical before time t_0.

2. *Time-invariance*. If an input function is shifted by a then the output is shifted by a. (An experiment at noon gives the same answer as an experiment at 2 p.m.)

3. *Linearity*. B is a linear map.

All systems discussed in the lectures will enjoy these three properties. An important property which some enjoy and some do not is:

4. *Stability*. Every decaying input function produces a decaying output function.

For concreteness the reader should think of the box as an electrical circuit.

Conceptually the box is what is given. It acts on some space of functions which we choose to put in the box and responds by putting out other functions. There is no a priori reason that the appropriate class of inputs should be L^2 or L^∞ or any other class. In fact which space we use depends on the problem we ultimately want to solve and which quantity we designate as the input or as the output. This choice is usually somewhat arbitrary; for example, on a circuit output with two terminals we usually measure the voltage $v(t)$ across the two terminals and the current $c(t)$ through the two wires. Whether we call $c(t)$ the input and $v(t)$ the output, or vice versa, is arbitrary. However, once we make a choice the general input-output theory we are describing holds.

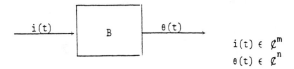

$i(t) \in \mathcal{L}^m$

$\theta(t) \in \mathcal{L}^n$

FIGURE 1.1. Black box.

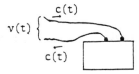

FIGURE 1.2.

Now we describe the mathematical setup we use in these chapters. We stick to single-input single-output systems (SISO in the literature), since generalization is straightforward. Let us suppose that all input functions lie in $L^2[0,\infty]$ and that our particular box B sends them to functions which grow no faster than e^{+Kt}; namely,

$$(1.1C) \qquad\qquad B \colon L^2[0,\infty] \to L^2[0,\infty;K]$$

where

$$L^2[0,\infty;K] \triangleq \{f \colon e^{-Kt}f(t) \text{ is in } L^2\}.$$

Note that (1.1C) implicitly uses *causality*. *Time invariance* says that B commutes with shifts

$$(1.1T) \qquad\qquad Bf_a = (Bf)_a \quad \text{for all } a > 0$$

where f_a stands for shifted f; that is, $f_a(t) = f(t-a)$. For the first few chapters *stability* will mean

$$(1.1S) \qquad\qquad B \colon L^2[0,\infty] \to L^2[0,\infty]$$

and *strictly stable* will mean that for some $\varepsilon > 0$ we have $B \colon L^2[0,\infty,-\varepsilon] \to L^2[0,\infty,-\varepsilon]$ and B is a bounded operator. Furthermore, the linear operators arising from physical systems will always map real-valued functions onto real-valued functions.

In studying a system it is very common to Laplace or Fourier transform all functions and work in "the frequency domain." Here the Laplace transform of a function f is

$$\tilde{f}(s) \triangleq \mathbf{L}f(s) = \int_0^\infty f(t)e^{-st}\, dt$$

and the usual Fourier transform is $\tilde{f}(i\omega)$, the Laplace transform $\tilde{f}$ restricted to the imaginary axis.

The inverse Laplace transform is

$$f(t) = \frac{1}{2\pi} \int_{k-i\infty}^{k+i\infty} f(s)\, ds$$

where the integral is taken over any contour $\operatorname{Re} s = K$ for which $f \in H^\infty(k + \text{R.H.P.})$.

Basic properties of the Laplace transform are

(1.2a) $$(\widetilde{e^{at}f})(s) = \tilde{f}(s-a),$$

(1.2b) $$\tilde{f}_a(s) = e^{-as}\tilde{f}(s), \qquad a > 0.$$

The Paley-Wiener theorem tells us that the Laplace transform $\tilde{f}$ of any function f in $L^2[0, \infty]$ extends analytically to all of the R.H.P., has values a.e. on the imaginary axis, and that the resulting function $\tilde{f}(i\omega)$ is in $L^2[-i\infty, i\infty]$. In fact $\tilde{L}^2[0, \infty] = H^2(\text{R.H.P.})$ where

$$H^2(\text{R.H.P.}) \triangleq \left\{ \begin{array}{l} h \in L^2 \text{ (imaginary axis): analytic in the R.H.P.,} \\ \int_{-\infty}^{\infty} |h(i\omega + a)|^2 \, d\omega \le M < \infty \quad \text{for all } a > 0. \end{array} \right.$$

The Laplace transform of $L^2[0, \infty; K]$ is exactly H^2 of the half-plane $\operatorname{Re} s \ge K$, the shift of R.H.P. by K. This is easily derived from the basic facts (1.2).

Intuitively $\tilde{f}(i\omega)$ is the contribution of $e^{i\omega t}$ to the function $f(t)$. For example, if $f(t)$ is a sound wave, then $|\tilde{f}(i\omega)|^2$ measures how much power of the wave is concentrated in the $e^{i\omega t}$ component of f. A device to measure $\tilde{f}(i\omega)$ for a signal $f(t)$ is called a spectrum analyzer. They are commonplace and many millions of dollars change hands each year on spectrum analyzers for electric signals, mechanical vibrations, and even sound. To be convincing I brought one along for analyzing sound. It is cheap so it only measures $|\tilde{f}|^2$, the "power spectrum," and not the phase of $\tilde{f}$. The device displays the graph of $\log |\tilde{f}|^2$ on the front with lights. Since I'll leave the gizmo on, those of you whose eyes can follow rapidly bouncing lights will be able to follow the rest of the lecture via its (instantaneous) Fourier transform.

Laplace transforms greatly facilitate the treatment of boxes. Define the "Laplace-transformed" B by

$$\tilde{B}\tilde{i} = (\widetilde{Bi}) \quad \text{for all } i \text{ in } L^2[0, \infty].$$

Our discussion of Laplace transforms indicates that

$$\tilde{B} \colon H^2(\text{R.H.P.}) \to H^2(K + \text{R.H.P.})$$

and that time invariance is equivalent to

(1.3) $$\tilde{B}(e^{-as}\tilde{i}) = e^{-as}\tilde{B}\tilde{i} \quad \text{for all } i \in L^2[0, \infty]$$

for $a > 0$. This holds for all $\operatorname{Re} a \ge 0$ by analytic continuation. That is, $\tilde{B}$ commutes with multiplication by $e^{(i\omega+a)s}$ for all ω and $a > 0$. Many operator theorists would quickly guess the following fundamental fact:

THEOREM 1.1 (VARIANT OF [**FS**]). *Suppose $m = n = 1$. If B is linear, time invariant, and causal, i.e., satisfies (1.T) and (1.C), then there is a function $b(s)$ defined and analytic for $\operatorname{Re} s > K$ such that*

$$[\tilde{B}\tilde{i}](s) = b(s)\tilde{i}(s) \quad \text{for } \operatorname{Re} s > K.$$

If B is stable as in (1.S), then $K = 0$ and b is in H^∞(R.H.P.) with $\sup_\omega |b(i\omega)| = \|B\|$. For a system with n inputs and m outputs everything goes through as above with b an $m \times n$ matrix-valued function and $\tilde{i}(s)$ a $\mathbb{C}^n$-valued function. Also all b which occur in engineering are real on the real axis.[1]

Practically all functions b in engineering are meromorphic on all of $\mathbb{C}$, and most b are actually rational. Consequently the theorem gives a more conservative impression than necessary.

IDEA OF PROOF FOR $m = n = 1$. If 1 were in H^2, then we would take $b = \tilde{B}1$. This is because

$$\tilde{B}e^{-\alpha s} = e^{-\alpha s}(\tilde{B}1)(s) = e^{-\alpha s}b(s)$$

for $\text{Re}\,\alpha \geq 0$ and so $[\tilde{B}p](s) = p(s)b(s)$ for any p which is a linear combination of the $e^{-\alpha s}$. These are dense in H^2. Now 1 is not in H^2, but if f is any function in H^2 formally $b = \widetilde{Bf}/\tilde{f}$. This can be used to give a proof.

ALTERNATIVE PROOF. Set $b = \widetilde{B\delta_0}$ where δ_a denotes the δ function supported at the point a. Use superposition. This has the physical interpretation that we send a sharp pulse δ_0 into the box. Measure the output $B\delta_0$, called the *impulse response function*, and Fourier transform it to get b.

The function b is called the *transfer function* or *frequency response function* (F.R.F.) of the system whose input-output operator is B. It has a simple physical interpretation befitting its name. For $m = n = 1$ and a strictly stable system, think of an input signal $\sin \omega t$ going into the system. To be more precise nothing is going in—we flip a switch to send in $\sin \omega t$. After a while the effect of the sudden transition dies out and what we observe *must be* of the form $A \sin(\omega t + \phi)$. This is always true for a strictly stable linear time invariant system. Algebraic manipulation is more convenient in complex notation; namely, think (fictionally) of sending a signal $e^{i\omega t}$ into the system—then $Ae^\phi e^{i\omega t}$ comes out. For a stable system *the frequency response function b could be defined at frequency ω to be the complex number* $\mathbf{A}e^{i\phi}$.

To "prove" this intuitively recall that the Fourier transform of $e^{i\omega t}$ is $\delta_{i\omega}$. Since the system is stable, by Theorem 1.1 b is defined on the $i\omega$ axis and

$$Be^{i\omega t} = \widecheck{b(s)\delta_{i\omega}} = b(i\omega)\check{\delta}_{i\omega} = b(i\omega)e^{i\omega t}.$$

This is not exactly correct since here we use $e^{i\omega t}$ for all $-\infty < t < \infty$. Thus our formula ignores the effect of the transition from 0 to $e^{i\omega t}$ which the Laplace transform method accounts for.

A more complete "proof" is also instructive (but can be skipped). The Laplace transform of an input $e^{i\omega t}$ is

$$\mathbf{L}e^{i\omega t} = \int_0^\infty e^{i\omega t - ts}\, dt = \frac{1}{i\omega - s}e^{(i\omega - s)t}\big\rvert^\infty = \frac{1}{s - i\omega} \quad \text{for } \text{Re}\,s \geq 0$$

---

[1] B always maps real valued functions to real valued functions.

and of the output is

$$\mathbf{L}(Be^{i\omega t}) = b(s)(\mathbf{L}e^{i\omega t}) = \frac{b(s)}{i\omega - s}.$$

Apply the inverse Laplace transform to get

$$[Be^{i\omega t}](\tau) = \frac{1}{2\pi}\int_{K-i\infty}^{K+i\infty}\frac{b(s)}{s - i\omega}e^{+s\tau}\,ds.$$

As an example take $b(s) = \beta/(s + c)$. Then

$$Be^{i\omega t} = \frac{\beta}{i\omega + c}[e^{i\omega t} - e^{-ct}] = b(i\omega)[e^{i\omega t} - e^{-ct}].$$

If $b \in H^{\infty}(\text{R.H.P.})$, that is, $c > 0$, then $e^{-ct} \to 0$ quickly, so indeed after a "short" time the output looks like $b(i\omega)e^{i\omega t}$ as advertised. If $c > 0$, then the output has an exponentially exploding part (which is consistent with the fact that b is unstable).

Note from the same computation that input $e^{(i\omega + r)t}$ gives output

$$Be^{(i\omega + r)t} = b(i\omega + r)[e^{(i\omega + r)t} + e^{-ct}].$$

Thus $b(i\omega + r)$ describes how B handles a complex frequency $i\omega + r$.

EXAMPLE 1.

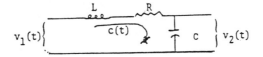

FIGURE 1.3.

Question. What v_2 is produced by v_1? That is, take v_1 as input, v_2 as output, and find the transfer function.

Solution. Let $c(t)$ be the current in the loop at time t. Kirchoff's Laws and the definition of inductor-capacitor, etc., give

$$L\frac{dc}{dt} + Rc + \frac{1}{C}\int c\,dt = v_1, \qquad \frac{1}{C}\int c\,dt = v_2.$$

Assume zero initial conditions and take Laplace transforms to get

$$Ls\tilde{c}(s) + R\tilde{c}(s) + \frac{1}{C}\frac{1}{s}\tilde{c}(s) = \tilde{v}_1(s), \qquad \frac{1}{C}\frac{1}{s}\tilde{c}(s) = \tilde{v}_2(s).$$

Eliminate $\tilde{c}(s)$ to get that the desired transfer function

$$\frac{\tilde{v}_2(s)}{\tilde{v}_1(s)} = \frac{1}{LCs^2 + RCs + 1}.$$

EXAMPLE 2 (see Figure 1.4).

This is a mechanical system driven by force $F(t)$ whose height $y(t)$ we want to measure. Take $F(t)$ as input, $y(t)$ as output. Then

$$M\frac{d^2y}{dt^2} + \phi\frac{dy}{dt} + ky = F.$$

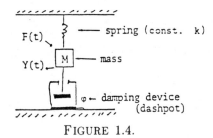

FIGURE 1.4.

If initially $y(0) = 0 = dy(0)/dt$, then the Laplace transform

$$\frac{\tilde{y}(s)}{\tilde{F}(s)} = \frac{1}{Ms^2 + \phi s + k}.$$

EXAMPLE 3. A typical model for a linear time invariant system is a box which has inputs i, outputs θ, and internal states x. The box is defined by linear maps A, B, C, D which relate states to inputs and outputs by

$$dx(t)/dt = Ax(t) + Bi(t), \qquad x(0) = 0, \qquad \theta(t) = Cx(t) + Di(t).$$

In Examples 1 and 2 the systems of differential equations can be written in this first order form. After Laplace transforming, the $i \to \theta$ transfer function is gotten from

$$s\tilde{x}(s) - A\tilde{x}(s) = B\tilde{i}(s), \qquad \tilde{\theta}(s) = C\tilde{x}(s) + D\tilde{i}(s)$$

and is

$$T(s) = D + C(sI - A)^{-1}B.$$

EXAMPLE 4. A BOEING 747 (BEFORE THE CONTROL SYSTEM IS ADDED).[2]

x, y, z	position coordinates	q	pitch rate
u, β, w	velocity coordinates	r	yaw rate
p	roll rate		

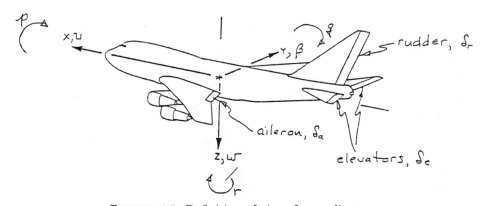

FIGURE 1.5. Definition of aircraft coordinates.

[2]Franklin, *Feedback control of dynamic systems*, Addison-Wesley, Reading, Mass., 1986, p. 475, Figure 7.29. Reprinted with permission.

The equations of motion are nonlinear and so one typically linearizes the equations about a fixed solution. For example, one solution to the equations corresponds to horizontal flight at 40,000 ft and 774 ft/sec and the equations for small lateral perturbations from this trajectory are

(1.4)
$$\begin{bmatrix} \dot{\beta} \\ \dot{r} \\ \dot{p} \\ \dot{q} \end{bmatrix} = \begin{bmatrix} -.0558 & -.9968 & .0802 & .0415 \\ .598 & -.115 & -.0318 & 0 \\ -3.05 & .388 & -.4650 & 0 \\ 0 & .0805 & 1 & 0 \end{bmatrix} \begin{bmatrix} \beta \\ r \\ p \\ q \end{bmatrix} + \begin{pmatrix} .0073 \\ -.175 \\ .153 \\ 0 \end{pmatrix} \delta_r,$$

$$y = [0 \ 1 \ 0 \ 0] \begin{bmatrix} \beta \\ r \\ p \\ q \end{bmatrix}.$$

The $\delta_r \to r$ transfer function is

(1.5) $$G(s) = \frac{r(s)}{\delta_r(s)} = \frac{-.475(s + .498)(s + .012 \pm j.488)}{(s + .0073)(s + .563)(s + .033 \pm j.947)}$$

so that the system has two stable real poles and a pair of stable complex poles referred to as the "*Dutch roll*." The stable real poles are referred to as the *spiral* mode ($p_1 = -.0073$) and the *roll* mode ($s_2 = -.563$). From looking at the natural roots, we see that the offending mode that needs repair for good pilot handling is the Dutch roll or the roots at $s = -.033 \pm j.95$. The roots have an acceptable frequency but their damping of $\varsigma \simeq .03$ is far short[3] of the desired $\varsigma \simeq .5$.

Connecting systems. A system with frequency response function P is often associated with the diagram

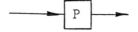

A common connection of two systems is

and its transfer function is QP so it is equivalent to

---

[3]This says that $G(s)$ has poles too close to the R.H.P. The $-.0073$ pole does not matter, because it has no imaginary component and therefore causes no oscillation, and the pilot can compensate for it easily himself.

There are junctions which add signals and junctions which subtract them.

The basic feedback system is

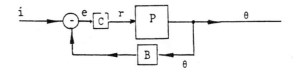

which is instructive to analyze. The diagram is equivalent to the equations:

$$\theta = PCe, \qquad e = i - B\theta.$$

Combine them to get $\theta = PCi - PCB\theta$. Consequently,

$$(I + PCB)\theta = PCi.$$

Then $\theta = (I + PCB)^{-1}PCi$ and so the transfer function for the system is

(1.6)[4] $$(I + PCB)^{-1}PC.$$

---

[4]This can be written in at least three other forms, with different dimensions of the inverse in nonsquare cases, by doing algebra in a different order. It involves $[I + PCB]^{-1}, [1 + BPC]^{-1},$ and $[I + CBP]^{-1}$.

2. Engineering Problems

This chapter begins by indicating in a general way what types of problems arise.

A basic type of optimization problem is this: Given $\Gamma(\omega, z)$ a positive-valued function of $\omega \in \mathbb{R}$ and of $z \in \mathbb{C}^N$, and given E a set of admissible functions on the imaginary axis, find

(OPT) $$\inf_{f \in E} \sup_{\omega} \Gamma(\omega, f(i\omega)).$$

While many different sets E occur, our attention will focus mostly on $E = H^\infty$(R.H.P.) or $E = \mathbf{A}$(R.H.P.) and the corresponding physical problem of designing systems to meet a stability constraint. Here $\mathbf{A}$ denotes the functions in H^∞ which are continuous on the closed half-plane including infinity.

Another basic type of problem is:

(REALIZE)   Given $f(s)$, build a system (using your technology, electronic, mechanical, etc.) whose transfer function equals $f(s)$.

Physically the problem is as follows: given the specs $f(i\omega)$, build a circuit or system with those specs. Frequently (OPT) and (REALIZE) are performed consecutively. First you find a "realizable" f which optimizes performance. Then you realize it as a circuit.

The chapters focus on (OPT), both its qualitative and quantitative aspects. However, the mathematics in Chapter 5 actually underlies one of the main areas in circuit realization which will get brief mention. Solving (OPT) is hard enough that one is usually delighted to find an $f \in E$ which for given c satisfies

(OPT) $$\sup_{\omega} \Gamma(\omega, f(i\omega)) \le c,$$

so we also call this problem (OPT) as well.

A fundamental point about (OPT) over $E = \mathbf{A}$ is that it can be reliably solved when the sublevel sets $\mathbf{S}_\omega(c) = \{z : \Gamma(\omega, z) < c\}$ are "disks." For z in $\mathbb{C}$ disk means the usual thing and (OPT) becomes: Does the function space "disk"

(2.1) $$\Delta_K^R \triangleq \{f : |K(i\omega) - f(i\omega)| < R(i\omega)\}$$

contain a function f in A? If so, compute f.

11

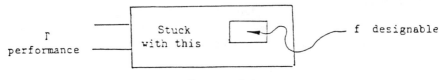

FIGURE 2.1.

THEOREM 2.1 (SEE D IN CHAPTER 5.D). *The intersection $A \cap \Delta_K^R$ can be computed explicitly, even when K, f, and R are matrix-valued functions.*

This theorem is a major topic in Chapters 3, 4, 5, and 6.

There is substantial motivation for (OPT) from engineering. Imagine you are designing a circuit or a system which is supposed to minimize a given performance measure at each frequency. You get to design part of the system, part is already forced on you. You are stuck with it (see Figure 2.1). At frequency ω, the performance penalty is some function $\Gamma(\omega, f(\omega))$. It depends on ω and what $f(i\omega)$ you choose for your design. The problem is as follows: Find

$$\inf_{f \text{ admis.}} \sup_{\omega} \Gamma(\omega, f(i\omega)),$$

where f sweeps those functions admissible to your design. I'll stick to stable circuits, that is, to f analytic in the R.H.P.

You might imagine from the generality of the definition that this problem comes up in many branches of engineering, and I think it is central to them. In fact there are several areas I know in which a miracle occurs. It is not a miracle of mathematics but one of science.

Miracle. The basic undergraduate textbook performance measures in amplifier design and control give Γ's whose sublevel sets $\mathbf{S}_\omega(c) = \{z : \Gamma(\omega, z) < c\}$ are *disks*.

This may occur in other branches of engineering too, but at least it is now appreciated that for amplifiers, things like gain, noise figure, and sensitivity, and for control systems, tracking error, gain-phase margin, bandwidth constraints, are disk constraints.

This talk gives two simple illustrations of how (OPT) arises; the first from control, the second from circuits. The control example is timely since extensions of it are inspiring substantial engineering interest.

EXAMPLE 1. Design an autopilot for your car (or what you do when you drive a car). Suppose the objective is to stay a given distance $d(t)$ from the yellow line (see Figure 2.2). The machine is not designed with a fixed $d(t)$ in mind but should respond well to any assignment $d(t)$. Assume the speed is constant, so we never have to worry about the accelerator or brakes. Let $\phi(t)$ be the angle of the steering wheel at time t. You as the driver observe your true distance $D(t)$ from the yellow line. What rule do you follow or should your autopilot follow to produce $\phi(t)$ to track $d(t)$?

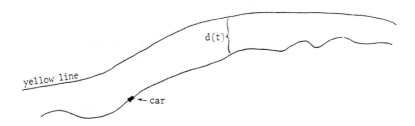

FIGURE 2.2.

Solution. Such a problem is always associated with a diagram as shown in Figure 2.3. The $\rightarrow \ominus$ takes one signal and subtracts the other; thus $e(t) = d(t) - D(t)$, which is the error your car is making at time t. The function $P(s)$ is the transfer function for your car for $\tilde{\phi}$ to $\tilde{D}$ at your fixed speed of travel. The slushiness or tautness of the steering in your car is captured in $P(s)$. $C(s)$ is the function you want to find, namely, your own transfer function.

Mathematical formulation. By (1.4) the $\tilde{d}$ to $\tilde{D}$ response function is $T \triangleq PC(1 + PC)^{-1}$. We can use T to parametrize all possible compensators since $C = P^{-1}T(1 - T)^{-1}$. Key constraints on the problem in terms of T are the following:

(1) For T to track d well; that is, for $\tilde{d} - \tilde{D} = (1 - T)\tilde{d}$ to be near zero we would like for $T(i\omega)$ to be near 1 for all ω. $T \cong 1$.

(2) T is stable. $T \in \mathbf{A}(\text{R.H.P.})$.

(3) T is not too close to[1] instability.   For a given M require $\|T\|_{L^\infty} \le M < \infty$.

(4) At frequencies higher than some fixed ω_B the total[2] system is nearly zero.   For a given $\varepsilon(\omega)$ and all $|\omega| > \omega_B$ we require $|T(i\omega)| < \varepsilon(\omega)$.

There are more constraints one might like, but these illustrate the situation. Some physical justification for (1), (2), and (3) has already been given, while item (4) is new. Item (4) is called a bandwidth constraint on the system. The reasons for it are varied: often one does not trust his computation of $P(s)$ at higher frequencies, so the safest thing is just to kill the system; the driver has a limited performance, e.g., most people cannot turn the wheel at the speed of light (in fact past a certain point the faster you turn the wheel the less arc you

---

[1] Engineers, this implies the constraint that the gain-phase margin be $\ge 1/(M + 1)$ for all ω.

[2] In engineering jargon this is called the closed loop system.

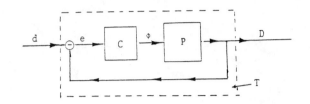

FIGURE 2.3.

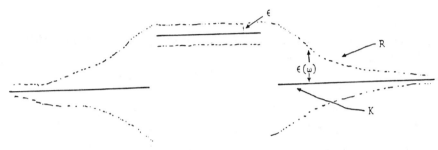

FIGURE 2.4.

can turn it through). One weakness of most undergraduate control texts is that they do not take (4) seriously. Engineers argue over how $\varepsilon(\omega)$ should be chosen.

The first observation in solving (1), (2), (3), and (4) is that (1) and (4) are completely contradictory. Fortunately in (1), one only needs $T(i\omega) \approx 1$ for low frequency ω. That is because tracking little squiggly variations in $d(t)$ is not important to driving. At last the mathematical problem is as follows:

Find T in $\mathbf{A}$ which also lies in the function space disk Δ_K^R where

(2.2)
$$K(i\omega) = \begin{cases} 1, & |\omega| < \omega_B, \\ 0, & |\omega| > \omega_B, \end{cases}$$

$$R(i\omega) = \begin{cases} \varepsilon_T, & |\omega| < \omega_B, \\ \varepsilon(\omega), & |\omega| \geq \omega_B. \end{cases}$$

Here we assume that $M > \max\{1+\varepsilon, \varepsilon(\omega)\}$ so that (3) is a redundant constraint and we can drop it. ε will be called the tracking error.

While the function K is a step function, in practice one is welcome to smooth out the jump substantially to facilitate computations. So a reasonable mathematical formulation of the basic control problem is: does there exist a function $f \in \mathbf{A}$ such that

(2.3) $|K(i\omega) - f(i\omega)| < R(i\omega)$

where K and R are smoothed versions of (2.2)? If so, find one. This is a "disk" problem which by Theorem 2.1 is very reliably soluble on the computer (provided the jump in K is sufficiently smoothed out; see Figure 2.4).

One generalization is to a system P with several inputs and outputs. Mathematically this corresponds to a matrix-valued P. Considerations above generalize in a straightforward way to matrix-valued functions; a thorough treatment of this requires the full weight of modern operator-theoretic approximation theory (Chapter 5).

There are other engineering restrictions which one would like to impose. One is called internal stability, and it causes no essentially new mathematical difficulty. Other manageable restrictions are discussed in [**FO**], [**FZ**], [**Doy2**], [**H8–H9**], and [**KT**].

A large class of engineering constraints has not been analyzed. These lead to open mathematics questions. First we have time domain restrictions; for example, a typical overshoot restriction is that the final system (called the closed loop system) maps the unit step function to a function less than 1.1. (Change from the outside to inside lane quickly without crossing the yellow line.) The mathematical question is:

OPEN QUESTION 2.1. Given Δ_K^R find a $T \in \Delta_K^R \cap \mathbf{A}$, real on the real axis, such that

$$\text{C.P.V.} \cdot \frac{1}{2\pi} \int_{-\infty}^{\infty} \frac{T(i\omega)}{i\omega} e^{i\omega t} d\omega \leq 1.1.$$

Note the integral is real-valued since $T(-i\omega) = \overline{T}(i\omega)$. Later we state an analog on the disk of Q2.1 which might be more appealing. Also we state mathematical formulations of other time domain constraints.

In control system design objective functions Γ which have sublevel sets more complicated than disks arise in several ways:

(1) Competing disk constraints yield functions Γ whose sublevel sets are intersections of disks.

(2) Throughout this discussion we have assumed that the transfer function $P(s)$ is known exactly before you start the design. Usually knowledge of $P(s)$ is imperfect. Accounting for this gives extremely general objective functions.

EXAMPLE 2. Circuit examples are in a way simpler to set up. Many simple circuit problems very quickly present one with well-posed hard (OPT) problems, while how to set up a simple control problem (as we just did) is something engineers still argue about.

Here is the simple gain problem sketchily described (see Figure 2.5). For complete definitions and details and more problems see [**BH8**] and [**H1–H7**]. One has a power source consisting of an ideal voltage source whose internal impedance is frequency dependent. We tie it to a frequency dependent load. Suppose S, the characteristic of the source, is given and the load is designable. Which choice of passive load L maximizes the power transfer ratio to the load?

Mathematical Problems. Given S in $\mathbf{A}$ with $\|S\|_{L^\infty} < 1$, the power transfer ratio to L at frequency ω is $1 - \Gamma(\omega, L(i\omega))$ where

$$(2.4) \qquad \Gamma(\omega, L(i\omega)) = \left| \frac{\overline{S}(i\omega) - L(i\omega)}{1 - S(i\omega)L(i\omega)} \right|^2.$$

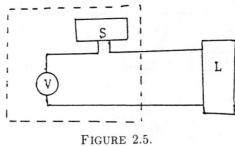

FIGURE 2.5.

Our problem is to solve (OPT) with E equal to all L in $\mathbf{A}$ with $\|L\|_{L^\infty} < 1$. The sublevel sets of Γ are clearly disks and so this is solved by Theorem 2.1.

Appendix to Part I: Historical Remarks

The Laplace transform and the frequency response function are too basic to be the attention of our discussion here. The approach to control and circuits described in Chapter 2 is surprisingly recent.

The theory of control is often split into two categories called classical control and LQG control. Classical control was developed at MIT Radiation Laboratories during WWII by a group including James, Nicols, and R. S. Phillips. They adapted techniques from Bode's method for designing amplifiers to control design. Mathematically the main technique is adjusting parameters in a low order (e.g., two) rational function while watching their graphs to see if they lie in certain regions. Modern control originated in the early 1960s with work of Kalman. Industry used classical control almost exclusively until recently when manipulation of matrix-valued functions became essential. Such problems quickly became too large for "parameter adjusting."

The techniques of Nevanlinna-Pick interpolation had their first serious introduction into engineering in a circuits paper of Youla and Sato [**YS**] in the middle 1960s. They were not pursued, and in the mid-seventies Helton [**H1–H3**] applied them to power transfer problems (like Chapter 2, Example 2). This was done at substantial generality in that the methods of commutant lifting [**S1–S2**], [**NF**], and Adamajan-Arov-Krein [**AAK1–AAK5**] were used to solve a many-input, many-output power transfer problem.

In the later 1970s G. Zames [**Z**] began to marshal arguments indicating that H^∞ rather than H^2 was the physically proper setting for control. He formulated the control problem much as was done here except without bandwidth type constraints. Within a few years Zames and Francis [**ZF**] used the Nevanlinna-Pick theory to solve the resulting problem and soon thereafter Francis-Helton-Zames solved it for many-input, many-output systems [**FHZ**]. Zames speculated widely that these methods were the appropriate ones for codifying classical control. By 1983 bandwidth constraints were successfully analyzed in independent efforts by Doyle [**Doy2**], Helton [**H8**], and Kwakernaak [**Kw**]. Since the original [**ZF**] paper there have been many extensions and improvements; in fact, they have been too numerous for us to list here. A few will appear in Chapters 7 and 10. A forthcoming survey by Doyle and Francis should have a complete bibliography.

Independent of Francis and Zames, Tannenbaum [**Ta**] in 1981 used interpolation techniques to solve a control problem. The approach is different than the one described here, and has led to many fruitful studies.

An excellent reference is the survey article by Bruce Francis and John Doyle [**FD**] or Francis's book [**Fr**].

Part II. Analytic Functions and Matrices

The goal is to describe a unified framework for obtaining a large number of results in analysis and matrix theory. It is quite surprising that so many different things follow from one mathematical core. The list of particulars include theories of the following:

1. Classical interpolation, H^∞ approximation, Corona-Bezout identities, commutant lifting.

2. Wiener Hopf, inner-outer, coprime factorization, Riemann-Hilbert problems.

3. Matrix LU decompositions, upper triangular matrix approximation matrix completion and interpolation.

4. Integrable systems (Toda lattice, mKdV).

5. The areas above can be refined to respect a substantial class of symmetries. Items 1, 2, and 5 incidentally constitute a fairly complete list of the main analysis tools available for frequency domain design of circuits and systems (when only input-output information rather than state space is used).

As we shall see when functions are smooth (or rational) most results in these areas can be proved from:

(1) a Beurling-Lax-Halmos type of representation for subspaces of L_n^2 which are invariant under multiplication by $e^{i\theta}$,

(2) basic facts about projective space.

Thus we get unified and simple proofs of most known theorems once one learns a few elementary properties of projective space. It is rather surprising that essentially all of the analysis in these theorems is contained in item (1), the rather easy invariant subspace representation. This approach is due to Joe Ball and myself [**BH1–BH7**], and all new results presented in this section were obtained jointly.

Chapter 3 teaches projective space, while Chapter 4 does invariant subspace representations. Chapters 5 and 6 prove all of the results central in the areas above (except for the Riemann-Hilbert problem).

Matrix and nest algebra analogs of the function space approach presented in Chapters 4, 5, and 6 were derived by Ball and Gohberg. Chapter 8 describes how the Ball-Gohberg, the Ball-Helton, and the Segal-Wilson theories all fit into one framework.

3. Fractional Maps and Grassmannians

The mathematical study of systems is rife with linear fractional maps both of the usual and of very general sorts. This is illustrated by the system connection examples at the end of Chapter 1. Linear fractional maps stem from the fact that series connections add numbers and parallel connections add reciprocals. These are both linear fractional operations and all circuit connections can be built from them.

This section consists of the following six parts:

A. System connections.

B. The Grassmannian viewpoint of L.F.T.'s; the scalar case.

C. The matrix and operator case.

D. Cross ratios.

E. Physical interpretation.

F. Homework.

A. System connections. The most general circuit connection behaves simply and elegantly. Think for the moment of a "linear" box which takes in a vector

$$\mathbf{i} = \begin{pmatrix} i_1 \\ \vdots \\ i_n \end{pmatrix},$$

and puts out a vector

$$\mathbf{o} = \begin{pmatrix} o_1 \\ \vdots \\ o_m \end{pmatrix};$$

it is described by the linear mapping, $M\mathbf{i} = \mathbf{o}$. Suppose that we want to connect two boxes in order to obtain a new one; to make thing simple we consider the case where one box $M = \begin{pmatrix} a & b \\ c & d \end{pmatrix}$ has two inputs and two outputs, the other w has a single input and output, and we connect the boxes as in Figure 3.1 to obtain q, a single-input, single-output system.

What is q in terms of M and w? The definitions of M and w say

$$ai_1 + bi_2 = o_1, \qquad ci_1 + di_2 = o_2, \qquad wi = o,$$

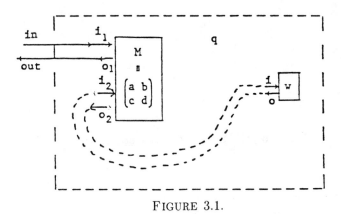

FIGURE 3.1.

while connecting w to M dictates $o_2 = i$ and $i_2 = o$. Substitution gives

$$ai_1 + bwi = o_1, \qquad ci_1 + dwi = i.$$

Use the second equation to eliminate i from the first equation and obtain $o_1 = \mathbf{F}_M(w)i_1,$ where $\mathbf{F}_M(w)$ is defined to be

$$(3.1) \qquad \mathbf{F}_M(w) = a + bw(1 - dw)^{-1}c.$$

Thus $q = \mathbf{F}_M(w)$, and one can think of this type of network connection (called *cascade connection*) as the action of a linear fractional map $\mathbf{F}_M$ on the number w. There is no philosophical reason to think of w as varying and a, b, c, d as fixed; however, this is a good mnemonic device. Obviously formula (3.1) is fractional linear in a, b, c, d, or w if everything else is kept fixed.

In this computation nothing keeps us from taking a, b, c, d, and w to be matrices or operators; only Figure 3.1 acquires more arrows. In fact if we take $a \in M_{lk}$, $b \in M_{lt}$, $c \in M_{rk}$, $d \in M_{rt}$, and $w \in M_{tr}$, then (2.2) produces an $\mathbf{F}_M(w)$ in M_{lk}, at least for generic w.

B. Linear fractional transformations: Scalar case. A more common form for fractional linear maps is

$$(3.2) \qquad \mathbf{G}_g(w) = (\alpha w + \beta)(\kappa w + \gamma)^{-1},$$

where g denotes $\left(\begin{smallmatrix} \alpha & \beta \\ \kappa & \gamma \end{smallmatrix}\right)$. Again this makes perfect sense for matrix entries provided $(\kappa w + \gamma)^{-1}$ is invertible. Conversion from (3.1) to (3.2) and vice versa is straightforward algebra and can be done in the absence of degeneracies.

The most powerful approach to studying linear fractional transformations (L.F.T.'s) is as linear mappings on a Grassmannian of subspaces. While most mathematicians are familiar with the idea, a repetition of it is not time consuming.

Think first of a conventional, linear fractional transformation $\mathbf{G}_g$ acting on scalars. Here g is an invertible 2×2 matrix with complex entries; that is, g is in $\mathrm{GL}(2, \mathbb{C})$. Instead of using the nonlinear formula $\mathbf{G}_g(w) = (\alpha w + \beta)(\kappa w + \gamma)^{-1}$, one could work with the matrix $g = \left(\begin{smallmatrix} \alpha & \beta \\ \kappa & \gamma \end{smallmatrix}\right)$. Obviously g acts on $\mathbb{C}^2$. A promising sign is that composition $\mathbf{G}_{g_1} \circ \mathbf{G}_{g_2}$ of L.F.T.'s equals $\mathbf{G}_{g_1 g_2}$ and so corresponds to matrix multiplication $g_1 g_2$. To describe the action of $\mathbf{G}_g$ on the complex plane in terms of the action of g on $\mathbb{C}^2$, one puts $\mathbb{C}$ into correspondence with one-dimensional subspaces of $\mathbb{C}^2$ as follows:

To w in $\mathbb{C}$ associate the space

$$\mathbf{S}(w) = \left\{ \begin{pmatrix} wz \\ z \end{pmatrix} : z \in \mathbb{C} \right\}$$

called the *graph* of w. This clearly gives a one-to-one correspondence between $\mathbb{C}$ and one-dimensional suspaces of $\mathbb{C}^2$ (excluding $\left\{ \begin{pmatrix} z \\ 0 \end{pmatrix} : z \in \mathbb{C} \right\}$). Common terminology dubs w the *angle operator* for $\mathbf{S}(w)$.

Since $g \colon \mathbb{C}^2 \to \mathbb{C}^2$ it certainly maps a one-dimensional subspace $\mathbf{S}$ to another one $\dot{\mathbf{S}}$. One easily checks that the effect on angle operators is

$$(3.3) \qquad\qquad g\mathbf{S}(w) = \mathbf{S}(\dot{w}) \Leftrightarrow \dot{w} = \mathbf{G}_g(w)$$

by the computation

$$g \begin{pmatrix} wz \\ z \end{pmatrix} = \begin{pmatrix} \alpha & \kappa \\ \beta & \gamma \end{pmatrix} \begin{pmatrix} wz \\ z \end{pmatrix} = \begin{pmatrix} [\alpha w + \kappa]z \\ [\beta w + \gamma]z \end{pmatrix}$$

$$= \begin{pmatrix} [\alpha w + \kappa][\beta w + \gamma]^{-1} y \\ y \end{pmatrix} = \begin{pmatrix} \mathbf{G}_g(w)y \\ y \end{pmatrix} \qquad \text{for some } y \in \mathbb{C}.$$

This computation is valid for all but one w.

A fundamental feature of *linear fractional maps* is that they *map circles to circles* and consequently disks to disks or their exterior (a half-plane is called a disk). This fact takes an elegant form in terms of g. Let Y be a selfadjoint matrix $\left(\begin{smallmatrix} a & f \\ \overline{f} & d \end{smallmatrix}\right)$. A disk Δ_Y in $\mathbb{C}$ is described algebraically as

$$(3.4) \qquad\qquad \{w \colon a|w|^2 + \overline{f}w + f\overline{w} + d \leq 0\}.$$

To give its Grassmannian equivalent, we associate the sesquilinear form

$$(3.5) \qquad\qquad \left(\begin{pmatrix} a & f \\ \overline{f} & d \end{pmatrix} \begin{pmatrix} u_1 \\ u_2 \end{pmatrix}, \begin{pmatrix} v_1 \\ v_2 \end{pmatrix} \right) \triangleq [u, v]_Y$$

on $\mathbb{C}^2$. A *subspace* $\mathbf{S}$ of $\mathbb{C}^2$ is called *negative* (resp. *null*) for a sesquilinear form $[\ ,\ ]$ provided that $[u, u] \leq 0$ (resp. $= 0$) for all $u \in \mathbf{S}$. Then $\mathbf{S}(w)$ is negative provided that

$$0 \geq \left(\begin{pmatrix} a & f \\ \overline{f} & d \end{pmatrix} \begin{pmatrix} wz \\ z \end{pmatrix}, \begin{pmatrix} wz \\ z \end{pmatrix} \right) = (a|w|^2 + \overline{f}w + \overline{w}f + d)|z|^2,$$

which, of course, is the same as w being in Δ_Y. That $\mathbf{G}_g$ maps disks to disks merely amounts to the fact that if $[u, v]_Y$ is a sesquilinear form, then $[gu, gv]_Y = [u, v]_{g^* Y g}$ is also. Thus $\mathbf{G}_g$ maps Δ_Y to $\Delta_{g^* Y g}$.

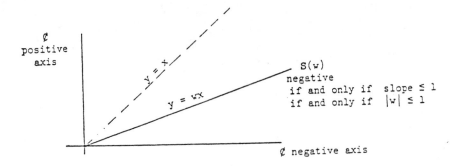

FIGURE 3.2. Negative spaces $\leftrightarrow$ unit disk in $\mathbb{C}$.

As an example take $Y = \left(\begin{smallmatrix} 1 & 0 \\ 0 & -1 \end{smallmatrix}\right)$. Then $\Delta_Y = \{w \colon |w| \leq 1\} =$ the unit disk and $\mathbf{S}(w)$ is negative in $[\ ,\ ]_Y$ if $|w| \leq 1$ (see Figure 3.2).

The maps $\mathbf{G}_g$ of the unit disk onto itself correspond to those g satisfying $g^* \left(\begin{smallmatrix} 1 & 0 \\ 0 & -1 \end{smallmatrix}\right) g = \left(\begin{smallmatrix} 1 & 0 \\ 0 & -1 \end{smallmatrix}\right)$.

C. The matrix and operator case. At this point we emphasize that frequently one can work with *either* g acting on subspaces or with $\mathbf{G}_g$ acting on $\mathbb{C}$; *you don't usually need to work with both.* Most of the complication above actually lies in going back and forth between the two viewpoints. Consequently there are many simple things one can do with matrices acting on a Grassmannian of subspaces, from which one gets a surprising amount of mileage.

Generalization of this to $\mathbb{C}^n$ and even to Hilbert space is very straightforward. To study matrix L.F.T.'s:

$$\mathbf{G}_g(w) = (\alpha w + \beta)(\kappa w + \gamma)^{-1},$$

we use the obvious generalizations of the scalar discussion. Here we have vector spaces $\mathbf{K}_1, \mathbf{K}_2, \dot{\mathbf{K}}_1, \dot{\mathbf{K}}_2$ and $\mathbf{G}_g$ maps $w \colon \mathbf{K}_2 \to \mathbf{K}_1$ to $\dot{w} \colon \dot{\mathbf{K}}_2 \to \dot{\mathbf{K}}_1$. For $\mathbf{G}_g$ to make sense we must have $\dim \mathbf{K}_2 = \dim \dot{\mathbf{K}}_2$. The coefficient matrix g maps
$$\mathbf{K} \overset{\triangle}{=} \overset{\mathbf{K}_1}{\underset{\mathbf{K}_2}{\oplus}} \text{ to } \overset{\dot{\mathbf{K}}_1}{\underset{\dot{\mathbf{K}}_2}{\oplus}} \text{ by}$$

$$\begin{pmatrix} \alpha & \beta \\ \kappa & \gamma \end{pmatrix} \colon \overset{\mathbf{K}_1}{\underset{\mathbf{K}_2}{\oplus}} \to \overset{\dot{\mathbf{K}}_1}{\underset{\dot{\mathbf{K}}_2}{\oplus}}.$$

With $w \colon \mathbf{K}_2 \to \mathbf{K}_1$ we associate its *graph*

$$\mathbf{S}(w) = \left\{ \begin{pmatrix} wz \\ z \end{pmatrix} \colon z \in \mathbf{K}_2 \right\}.$$

Conversely with a subspace whose dimension equals $\dim \mathbf{K}_2$, we associate an *angle operator* w provided the subspace contains no vector of the form $\left(\begin{smallmatrix} z \\ 0 \end{smallmatrix}\right)$. If $\dim \mathbf{S} < \dim \mathbf{K}_2$, one still could associate an "angle operator" w to $\mathbf{S}$, except w is only defined on a proper subspace $\mathbf{D}$ of $\mathbf{K}_2$. The space $\mathbf{D}$ will be the *domain* of the angle operator w for $\mathbf{S}$.

All basic facts from before stay about the same and we list them, while re-marking that all proofs go exactly as before. For w, defined on all of $\mathbf{K}_2$, we have

FACT 3.C.1.[1] $g\mathbf{S}(w) = \mathbf{S}(\dot{w}) \Leftrightarrow \dot{w} = \mathbf{G}_g(w)$.

For inner product spaces $\mathbf{K}_1$ and $\mathbf{K}_2$, a disk Δ_Y in the space of maps $w\colon \mathbf{K}_1 \to \mathbf{K}_2$ has the form

$$\Delta_Y = \{w\colon w^*aw + f^*w + w^*f + d \leq 0\}$$

where $Y = \left(\begin{smallmatrix} a & f \\ f^* & d \end{smallmatrix}\right)$ is selfadjoint. Again

FACT 3.C.2. $\mathbf{G}_g\colon \Delta_Y \to \Delta_{g^*Yg}$.

We shall need much more detail on spaces with sesquilinear forms $[\;,\;]$. A Krein space $\mathbf{K}\,[\;,\;]$ is defined to be a Hilbert space $\mathbf{K}$ with a "nondegenerate" sesquilinear form $[\;,\;]$. One can derive a property of *Krein space* which we actually prefer to use as a definition. Namely, $\mathbf{K}$ can be decomposed as the orthogonal direct sum of two Hilbert spaces $\mathbf{K}_1$ and $\mathbf{K}_2$ and with respect to this decomposition $[\;,\;] = [\;,\;]_Y$ where $Y = \left(\begin{smallmatrix} I & 0 \\ 0 & -I \end{smallmatrix}\right)$. Since $\mathbf{K}_1$ is $[\;,\;]$ strictly positive, and $\mathbf{K}_2$ is $[\;,\;]$ strictly negative, we henceforth call them $\mathbf{K}_+$ and $\mathbf{K}_-$ respectively; henceforth we denote $\left(\begin{smallmatrix} I & 0 \\ 0 & -I \end{smallmatrix}\right)$ by J. It is convenient to set up a notation for the Hilbert space inner products on $\mathbf{K}_\pm$ induced by $[\;,\;]$; so define

$$\langle u, v \rangle_+ \overset{\Delta}{=} \left[\begin{pmatrix} u \\ 0 \end{pmatrix}, \begin{pmatrix} v \\ 0 \end{pmatrix} \right] \quad \text{for } u, v \in \mathbf{K}_+,$$

$$\langle u, v \rangle_- \overset{\Delta}{=} -\left[\begin{pmatrix} 0 \\ u \end{pmatrix}, \begin{pmatrix} 0 \\ v \end{pmatrix} \right] \quad \text{for } u, v \in \mathbf{K}_-.$$

Then the Hilbert space inner product $\langle\;,\;\rangle$ on $\mathbf{K}$ is

$$\langle x, y \rangle = \langle x_+, y_+ \rangle_+ + \langle x_-, y_- \rangle_-,$$

while

$$[\;,\;] = \langle x_+, y_+ \rangle_+ - \langle x_-, y_- \rangle_-.$$

Now we list some standard geometric properties of a Krein space.

A subspace $\mathbf{S}$ is *negative, null,* or *positive* if $[u, u]$ is ≤ 0, $= 0$, or ≥ 0 for all $u \in \mathbf{S}$; *strictly negative,* of course, means $[u, u] < 0$, etc. A *maximal negative* space $\mathbf{S}$ is one which is contained in no other negative space. Two vectors x and y are called $[\;,\;]$-orthogonal if $[x, y] = 0$. Any subspace $\mathbf{M}$ has a closed $[\;,\;]$-*orthogonal complement* denoted $\mathbf{M}'$. These behave like orthogonal complements in a definite metric, except it is possible that $\mathbf{M} \cap \mathbf{M}' \overset{\Delta}{=} \gamma$ is not trivial. Clearly γ is a null space since any x in $\mathbf{M}$ is $[\;,\;]$-orthogonal to any y in $\mathbf{M}$ (including x). Note that $\mathbf{K}' = \{0\}$.

---

[1] This assumes that $g\mathbf{S}(w)$ is a subspace which possesses an angle operator.

If $\mathbf{M}$ and $\mathbf{N}$ are disjoint, $[\ ,\ ]$-orthogonal, and $\mathbf{M} + \mathbf{N}$ is closed, we denote their sum by $\mathbf{M} \boxplus \mathbf{N}$. Note that the sum of two negative spaces typically is not negative, while *the sum of* $[\ ,\ ]$-*orthogonal negative spaces is negative.*

Since our Krein space setting is a refinement of the one behind Facts 3.C.1, 3.A.2, etc., they hold for $\mathbf{K}_+$ and $\mathbf{K}_-$ and we also get

FACT 3.C.3. Every maximal negative subspace has an angle operator in $\Delta_J = \{w \colon \mathbf{K}_- \to \mathbf{K}_+ | w^* w \leq I\}$ and conversely. Negative subspaces correspond to $\{w \colon \mathbf{D} \subset \mathbf{K}_- \to \mathbf{K}_+ | w^* w \leq I\}$ where $\mathbf{D}$ is some subspace of $\mathbf{K}_-$.

The proof of this is an instructive exercise. It is clear from Fact 3.C.3 that a w in Δ_J produces a negative $\mathbf{S}(w)$. The space $\mathbf{S}(w)$ is not maximal negative if and only if there is a negative vector $x = x_+ + x_-$ not contained in $\mathbf{S}(w)$ which is $[\ ,\ ]$-orthogonal to $\mathbf{S}(w)$. That is

$$\left[\begin{pmatrix} wz \\ z \end{pmatrix}, \begin{pmatrix} x_+ \\ x_- \end{pmatrix} \right] = 0$$

for all $z \in \mathbf{K}_-$. Thus $\langle z, w^* x_+ - x_- \rangle_- = 0$ and so $w^* x_+ = x_-$. However, $0 \geq \|x_+\|^2 - \|x_-\|^2 = \|w^* x_1\|^2 - \|x_+\|^2$ which contradicts $\|w\| \leq 1$. Thus

$$w \in \Delta_J \Rightarrow \mathbf{S}(w) \text{ maximal negative.}$$

Conversely, $\mathbf{S}$ maximal negative implies that $\mathbf{S} \cap \mathbf{K}_+ = \{0\}$; so an angle operator $w \colon D \to \mathbf{K}_+$ exists on a subspace $\mathbf{D}$ of $\mathbf{K}_-$. If $\mathbf{D} \neq \mathbf{K}_-$, then pick x_- in $\mathbf{K}_-$ orthogonal to $\mathbf{D}$. The vector $\begin{pmatrix} 0 \\ x_- \end{pmatrix}$ is $[\ ,\ ]$-orthogonal to $\mathbf{S}$ and it is negative. This produces a contradiction.

A negative subspace of $\mathbf{N}$, which is maximal negative in $\mathbf{M}$, will be called $\mathbf{M}$-maximal negative; $\mathbf{N}$ need not be maximal negative in all of $\mathbf{K}$. By the *negative signature* of a subspace $\mathbf{M}$ of the Krein space $\mathbf{K}$ we mean the dimension l $(0 \leq l \leq \infty)$ of any maximal, strictly negative, subspace of $\mathbf{M}$. It turns out that this dimension is independent of which particular $\mathbf{M}$-maximal, strictly negative subspace one chooses, and thus negative signature is well-defined; if $\mathbf{M}$ contains no x which is $[\ ,\ ]$-orthogonal to all of $\mathbf{M}$, then this quantity is also the dimension of any $\mathbf{M}$-maximal negative subspace. The *negative cosignature* of the negative subspace $\mathbf{N}$ is the codimension of $\mathbf{N}$ inside of some maximal negative subspace $\mathbf{N}_1$ of $\mathbf{K}$; this quantity also is well-defined; that is, it is independent of the choice of maximal negative subspace $\mathbf{N}_1$ containing $\mathbf{N}$.

The Krein space, which will ultimately be of most interest to us, is in L_N^2, endowed with a sesquilinear form $[\ ,\ ]$, gotten from a sesquilinear form $[\ ,\ ]_{\mathbb{C}^N}$ on $\mathbb{C}^N$ by

$$[f, h] = \frac{1}{2\pi} \int_0^{2\pi} [f(e^{i\theta}), h(e^{i\theta})]_{\mathbb{C}^N} \, d\theta.$$

Unless otherwise stated take

$$[x, y]_{\mathbb{C}^N} = x_1 \overline{y}_1 + \cdots + x_m \overline{y}_m - x_{m+1} \overline{y}_{m+1} \cdots - x_{m+n} y_{m+n}.$$

In this case L_N^2, $[\ ,\ ]$ is a Krein space $\mathbf{K}$ with a decomposition $\mathbf{K}_+ = L_m^2$ and $\mathbf{K}_- = L_n^2$. Here $m + n = \mathbf{N}$. We will be concerned with subspaces $\mathbf{S} \subset L_N^2$ invariant under multiplication by $e^{i\theta}$; that is, $e^{i\theta}\mathbf{S} \subset \mathbf{S}$.

The most important fact for us about angle operators sits in this context, and the reader should try to understand it very thoroughly.

FACT 3.C.4. If $\mathbf{S}$ is an invariant maximal, negative subspace of H_N^2, then the angle operator F for $\mathbf{S}$ is multiplication by a function f in $\overline{\mathbf{B}}H_{m\times n}^\infty$, and conversely. Those $\mathbf{S}$ with negative cosignature l correspond to f in $\overline{\mathbf{B}}H_{m\times n}^{\infty;l}$.

PROOF. By Fact 3.C.3 any negative $\mathbf{S}$ has an angle operator $F\colon \mathbf{D} \to \mathbf{K}_+ = H_m^2$ with $\mathbf{D}$ a subspace of $\mathbf{K}_- = H_n^2$. Invariance of $\mathbf{S}$ implies $e^{i\theta}\mathbf{D} \subset \mathbf{D}$ and $e^{i\theta}F = Fe^{i\theta}$. If $\mathbf{S}$ is maximal negative, then $\mathbf{D} = \mathbf{K}_- = H_n^2$ and F is multiplication by a function f. In fact the jth column of f is $f\binom{0}{1}$; so $\mathbf{S} \subset H_N^2$ implies $f \in H_{m\times n}^\infty$.

For $\mathbf{S}$ not maximal negative l equals the codimension of $\mathbf{D}$ in $\mathbf{K}_-$. Consequently the classical Beurling-Lax theorem (Corollary 5.2) implies there exists an inner function ϕ with l zeros such that $\phi \mathbf{K}_- = \mathbf{D}$. Thus $F\phi\colon \mathbf{K}_- \to \mathbf{K}_+$ and $e^{i\theta}F\phi = F\phi e^{i\theta}$. As before, $F\phi$ is multiplication by a function f in $\overline{\mathbf{B}}H_{m\times n}^\infty$. As a result F is multiplication by $f\phi^{-1}$ which is in $\overline{\mathbf{B}}H_{m\times n}^{\infty,l}$.

The converse of all of these statements is obvious. Indeed the proof shows that a slightly more concrete way to present Fact 3.C.4 is to put $\mathbf{S}$ in correspondence with the pair $\binom{f}{\phi}$. In fact it is an easy (homework) exercise to prove the (eventually essential) variant on Fact 3.C.4:

FACT 3.C.4'. If $\mathbf{S} \subset \binom{L_m^2}{H_n^2}$ is negative with negative cosignature l, then it is

$$\mathbf{S} = \left\{ \begin{pmatrix} fu \\ \phi u \end{pmatrix} : u \in H_n^2 \right\}$$

where $\phi \in H_{n\times n}^\infty$ is inner and has degree l and $f \in \overline{\mathbf{B}}L_{m\times n}^\infty$.

Example.

(1) Take $f(z) = \frac{1}{2}z$, then $\mathbf{S}(f)$ is the space

$$\left\{ \begin{pmatrix} \frac{1}{2}zu(z) \\ u(z) \end{pmatrix} : u \in H^2 \right\}.$$

(2) Take

$$f(z) = \frac{1}{2}\frac{z}{z - \frac{1}{5}};$$

then $\mathbf{S}(f)$ is not

$$\left\{ \begin{pmatrix} \frac{1}{2}\dfrac{zu(z)}{z - \frac{1}{5}} \\ u(z) \end{pmatrix} : u \in H^2 \right\},$$

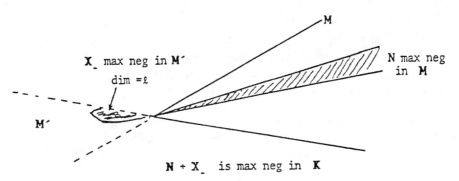

FIGURE 3.3. A picture of Fact 3.C.5 with $\mathbf{X}_0 = 0$.

since this is not contained in H_2^2. To f we associate the angle opertor F whose domain $\mathbf{D}$ is $\{u : fu \in H^2\} = \{u \in H^2 : u(\frac{1}{5}) = 0\}$. Then we take $\mathbf{S}(f)$ to be

$$\left\{ \begin{pmatrix} \frac{1}{2}(z - \frac{1}{5}) \\ u(z) \end{pmatrix} : u \in H^2 \right\}.$$

The following lemma on a general Krein space $\mathbf{K}$ is basic for our analysis of interpolation problems to come in Chapter 5.

FACT 3.C.5. Suppose $\mathbf{M}$ is a pseudoregular[2] subspace of $\mathbf{K}$. Then each $\mathbf{M}$-maximal negative subspace of $\mathbf{M}$ has negative cosignature l equal to the negative signature of $\mathbf{M}'$. (See Figure 3.3.)

PROOF OF FACT 3.C.5. It is not difficult to see that the space $\mathbf{X} \overset{\Delta}{=} \mathbf{M}'$ has a $[\ ,\ ]$-orthogonal decomposition

$$\mathbf{X} = \text{closure}(\mathbf{X}_+ \boxplus \mathbf{X}_- \boxplus \mathbf{X}_0)$$

into a strictly positive subspace $\mathbf{X}_+$, a strictly negative subspace $\mathbf{X}_-$, and a null space $\mathbf{X}_0$ $(= \mathbf{X} \cap \mathbf{X}' = \mathbf{M} \cap \mathbf{M}')$, where $\dim \mathbf{X}_-$ is the negative signature of $\mathbf{X}$. Suppose $\mathbf{N}$ is a $\mathbf{M}$-maximal negative subspace $(\mathbf{M} = \mathbf{X}')$; we claim that $\mathbf{N} + \mathbf{X}_-$

---

[2]There are some technical definitions which we have to introduce someplace. An engineer and many mathematicians can skip them since all spaces he will use are well behaved. The subspace $\mathbf{M}$ is called *nondegenerate* if no x in $\mathbf{M}$ is $[\ ,\ ]$-orthogonal to $\mathbf{M}$ (i.e., $\mathbf{M} \cap \mathbf{M}' = \{0\}$), and *regular* if there is no sequence $\{x_n\} \subset \mathbf{M}$ such that

$$\lim_{n \to \infty} \sup_{y \in \mathbf{M}} \frac{|[x_n, y]|}{\|x_n\| \, \|y\|} = 0.$$

That is, $\mathbf{M}$ is topologically nondegenerate.

FACT 3.C.6. $\mathbf{M}$ is nondegenerate if and only if $\mathbf{M} + \mathbf{M}'$ is dense in $\mathbf{K}$, and is regular if and only if in addition $\mathbf{M} + \mathbf{M}'$ is closed (and thus $\mathbf{K} = \mathbf{M} \boxplus \mathbf{M}'$). It is an easy corollary of [**Bog**, Theorem V, 3.5] that $\mathbf{M}'$ is regular (in our terminology) if and only if $\mathbf{M}$ is regular.

We say that the subspace $\mathbf{M}$ of a Krein space $\mathbf{K}$ is pseudoregular if $\mathbf{M} + \mathbf{M}'$ is closed. For an arbitrary subspace $\mathbf{M}$, it is always the case that $\mathbf{M} + \mathbf{M}'$ is dense in $(\mathbf{M} \cap \mathbf{M}')'$; thus $\mathbf{M}$ is pseudoregular if and only if we have the equality $\mathbf{M} + \mathbf{M}' = (\mathbf{M} \cap \mathbf{M}')'$. Clearly $\mathbf{M}$ is pseudoregular if and only if $\mathbf{M}'$ is pseudoregular.

is $\mathbf{K}$-maximal negative. From this it follows that any such $\mathbf{N}$ has codimension equal to $\dim \mathbf{X}_-$ in a $\mathbf{K}$-maximal negative subspace, and the lemma follows.

To prove that $\mathbf{N} + \mathbf{K}_-$ is maximal negative, we first observe that since $\mathbf{N}$ and $\mathbf{X}_-$ are $[\ ,\ ]$-orthogonal, $\mathbf{N} + \mathbf{K}_-$ is negative. Second, note that pseudoregularity implies $\mathbf{N} + \mathbf{X}_-$ is closed. To prove $\mathbf{N} + \mathbf{X}_-$ is maximal negative, we need only show that $(\mathbf{N} + \mathbf{X}_-)' = \mathbf{N}' \cap \mathbf{X}_-' = \mathbf{N}' \cap \text{clos}(\mathbf{M} + \mathbf{X}_+)$ is positive. This space equals closure$(\mathbf{X}_+ + \mathbf{N}' \cap \mathbf{M})$ because $\mathbf{N}' \supset \mathbf{X}_+$. Since $\mathbf{N}$ is $\mathbf{M}$-maximal negative, $\mathbf{N}' \cap \mathbf{M}$ is positive. By orthogonality and the positivity of $\mathbf{X}_+$, it next follows that $(\mathbf{N}' \cap \mathbf{M}) + \mathbf{X}_+$ is positive, as claimed.

D. Cross ratios. While it is something of an aside and inessential to these lectures, we mention cross ratios. First we treat the scalar case. To build invariants of fractional linear maps, traditionally one uses cross ratios of points w_j in $\mathbb{C}$:

$$\mathbf{C}(w_1, w_2, w_3, w_4) = (w_1 - w_2)(w_2 - w_3)^{-1}(w_3 - w_4)(w_4 - w_1)^{-1}.$$

The basic fact is that if $\mathbf{G}_g(w_j) = \dot{w}_j$, then

$$(3.6) \qquad \mathbf{C}(\dot{w}_1, \dot{w}_2, \dot{w}_3, \dot{w}_4) = \mathbf{C}(w_1, w_2, w_3, w_4).$$

The Grassmannian equivalent is the cross ratio $\mathbf{C}(\mathbf{S}_1, \mathbf{S}_2, \mathbf{S}_3, \mathbf{S}_4)$ of four (noncollinear) subspaces of $\mathbb{C}^2$ defined like this. Since $\mathbb{C}^2 = \mathbf{S}_1 + \mathbf{S}_3$, any x in $\mathbf{S}_2$ can be decomposed as a sum $x = x_1 + x_3$ of vectors in $\mathbf{S}_1$ and $\mathbf{S}_3$. This induces a map m from $\mathbf{S}_1$ to $\mathbf{S}_3$, via $mx_1 = x_3$. Likewise $\ddot{x}$ in $\mathbf{S}_4$ decomposes as $\ddot{x} = \ddot{x}_1 + \ddot{x}_3$ which induces a map p from $\mathbf{S}_3$ to $\mathbf{S}_1$ via $p\ddot{x}_3 = \ddot{x}_1$. We have $m \colon \mathbf{S}_1 \to \mathbf{S}_3$ and $p \colon \mathbf{S}_3 \to \mathbf{S}_1$, and we denote the composition map $pm \colon \mathbf{S}_1 \to \mathbf{S}_1$ by $\tilde{\mathbf{C}}(\mathbf{S}_1, \mathbf{S}_2, \mathbf{S}_3, \mathbf{S}_4)$. Since $\mathbf{S}_1$ is one-dimensional, this map corresponds to a $|x|$ matrix, that is, to a complex scalar, which, when $\mathbf{S}_j = \mathbf{S}(w_j)$, equals $\mathbf{C}(w_1, w_2, w_3, w_4)$. We remark that in this notation $pmx_1 = \ddot{x}_1$.

Invariance of the cross section $\tilde{\mathbf{C}}$ follows because g induces maps $\dot{m} \colon g\mathbf{S}_1 \to g\mathbf{S}_3$ and $\dot{p} \colon g\mathbf{S}_3 \to g\mathbf{S}_1$ given by

$$\dot{m}gx_1 = gx_3 \quad \text{since } gx = gx_1 + gx_3,$$

$$\dot{p}g\ddot{x}_3 = g\ddot{x}_1 \quad \text{since } g\ddot{x} = g\ddot{x}_1 + g\ddot{x}_3.$$

So

$$\dot{p}\dot{m}gx_1 = \dot{p}gx_3 = \dot{p}g\ddot{x}_3 = g\ddot{x}_1 = gpmx_1,$$

where a middle equality holds because $x_3 = \ddot{x}_3$. The identity

$$(3.7) \qquad \dot{p}\dot{m}g|_{\mathbf{S}_1} = gpm|_{\mathbf{S}_1},$$

when $\mathbf{S}_1$ is one-dimensional, which, if $\mathbf{S}_1$ and $g\mathbf{S}_1$ are identified, could be interpreted to say that the cross ratios are equal.

A final fact about cross ratios is that if two four-tuples w_1, w_2, w_3, w_4 and $\dot{w}_1, \dot{w}_2, \dot{w}_3, \dot{w}_4$ have the same cross ratio, there is a linear fraction transformation $\mathbf{G}_g$ which maps each w_j to $\dot{w}_j$. The map $\mathbf{G}_g$ is uniquely determined provided the w_j's are distinct.

The Grassmannian proof is very swift. Given two 4-tuples, $\mathbf{S}_j$ and $\dot{\mathbf{S}}_j$, we find a g with $g\mathbf{S}_j = \dot{\mathbf{S}}_j$. The cross ratios being equal as in (3.7) means precisely that there is an invertible map $g_0 \colon \mathbf{S}_1 \to \dot{\mathbf{S}}_1$ for which $\dot{p}'\dot{m}'g_0 = g_0 pm$ on $\mathbf{S}_1$. Define g by extending

$$g = \begin{cases} g_0 & \text{on } \mathbf{S}_1, \\ \dot{m}g_0 m^{-1} & \text{on } \mathbf{S}_3 \end{cases}$$

via linearity to all of $\mathbb{C}^2 = \mathbf{S}_1 + \mathbf{S}_3$. Clearly g is defined on all of $\mathbf{S}_1 + \mathbf{S}_3 = \mathbb{C}^2$. The point is to check that $g\mathbf{S}_j = \dot{\mathbf{S}}_j$ for $j = 2, 4$. If $x \in \mathbf{S}_2$, it equals $x_1 + mx_1$ and gx is $gx_1 + gmx_1 = g_0 x_1 + \dot{m}g_0 m^{-1}mx_1 = g_0 x_1 + \dot{m}g_0 x_1$, which (by the definition of $\dot{m}$) is in $\dot{\mathbf{S}}_2$. Thus $g \colon \mathbf{S}_2 \to \dot{\mathbf{S}}_2$. Likewise $g \colon \mathbf{S}_4 \to \dot{\mathbf{S}}_4$ and finite dimensionality forces these maps to be onto. Note that $\dot{m}$ and m are bijective since $\mathbf{S}_1$ and $\mathbf{S}_3$ are not collinear.

All of our scalar cross-ratio constructions were done in a way which makes extension to matrix cross ratios obvious. Our subspace construction generalizes without change to the case where $\mathbf{S}_1, \mathbf{S}_2, \mathbf{S}_3, \mathbf{S}_4$ all have the same finite dimension, do not intersect, and $\mathbf{S}_1 + \mathbf{S}_3$ spans $\mathbf{K}$. Now $\mathbf{C}$ is invariant up to similarity in a way which the following fact states precisely.

FACT 3.D.1. For $\dot{\mathbf{S}}_j = g\mathbf{S}_j$, the map $g_0 \overset{\Delta}{=} g|_{S_1}$ satisfies

$$(3.8) \qquad \tilde{\mathbf{C}}(\dot{\mathbf{S}}_1, \dot{\mathbf{S}}_2, \dot{\mathbf{S}}_3, \dot{\mathbf{S}}_4)g_0 = g_0\tilde{\mathbf{C}}(\mathbf{S}_1, \mathbf{S}_2, \mathbf{S}_3, \mathbf{S}_4).$$

This follows immediately from (3.7). Also one has the following:

FACT 3.D.2. If four $\mathbf{S}_j$ and $\dot{\mathbf{S}}_j$ as above have similar cross ratios, that is, $\mathbf{C}g_0 = g_0\tilde{\mathbf{C}}$, then there is an invertible g which maps $\mathbf{S}_j$ onto $\dot{\mathbf{S}}_j$ for all j.

The proof of this goes exactly as the proof we gave in the scalar case. The matrix cross ratio (not Grassmannian style) was introduced by the circuit theorist D. C. Youla in the 1950s, who learned of a special case from C. L. Siegel. Gelfand and Pomoronov did a Grassmannian version at great generality about 1970. Facts 3.D.1 and 3.D.2 are due to them respectively. An elegant algebraic treatment is in Schwartz and Zaks [**SZ1-SZ2**].

E. Physical interpretation. Philosophically from an engineering point of view Grassmannians actually are more natural than input-output operators M. Indeed, there is a senior level circuits book [**KR**] from the 1960s which takes this point of view. Practical engineers don't use the Grassmannian at all and pay the price of having to change coordinates very often. Many senior circuit books have a full chapter on these changes of coordinates.

To think of a simple linear circuit in Grassmannian terms (see Figure 3.4), we take our circuit C and hook it to a very versatile power generator. Record all volt current pairs $(v(t), i(t))$ which you measure. The set of all volt-current pairs which can occur is a subspace $\mathbf{S}$ of the set of "all pairs" of functions. The true external description of the circuit is exactly the space $\mathbf{S}$. Note there is no mention of inputs and outputs.

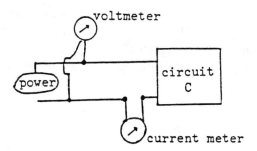

FIGURE 3.4.

If one decides to think of current as input and voltage as output, then one associates the operator Z with $\mathbf{S}$ via

$$\mathbf{S} = \{(Zi, i) \colon \text{all } i\}$$

and Z is called the *impedance operator* for C. If we elect to take voltage as input and current as output, then we get $Yv = i$ called the *admittance operator* for C. Likewise,

$$W\left(\frac{i+v}{2}\right) = \left(\frac{i-v}{2}\right)$$

is the *scattering operator*. There are also *hybrid operators* and *chain operators*.

All of this amounts to what a mathematician would call choosing various *affine coordinates* on a Grassmannian of subspaces. In our terms it amounts to selecting different decompositions of the space $\mathbf{K}$ into $\mathbf{K}_1 + \mathbf{K}_2$.

The astute reader will have observed that to some subspaces one can associate a Z, but not a Y or vice versa. This is one reason a designer will sometimes use Y over Z, or S, etc. (Another reason is ease of computation.) At any rate the dullest chapter in any circuit book is the one which tells you how to convert between Y's, Z's, S's, H's, etc.

Also natural to circuits are signed sesquilinear forms. This is because energy in electronics is a signed form *not a positive definite one*. In fact the power consumed by a circuit when voltage $v(t)$ and current $i(t)$ are measured is

$$P(i, v) = \int_0^\infty \operatorname{Re} v(t)\overline{i(t)}\, dt.$$

Consequently the signed form $[\,,\,]$ defined by

$$[(v_1, i_1), (v_2, i_2)] = \operatorname{Re}(v_1\overline{i_2} + v_2\overline{i_1})$$

plays a fundamental role. Formally we can express $[\,,\,]$ as

$$[(v_1, i_1), (v_2, i_2)] = \tfrac{1}{4}(\|v + i\|^2 - \|v - i\|^2).$$

Thus we set $\mathbf{K}_\pm = \{(v, i) \colon v = \pm i\}$ which are clearly positive and negative, and $\mathbf{K} = \mathbf{K}_+ + \mathbf{K}_-$ yields a Krein space. Each linear circuit C produces a subspace $\mathbf{S}$ of $\mathbf{K}$, and its angle operator with respect to $\mathbf{K}_+, \mathbf{K}_-$ is the scattering operator W referred to above.

A *passive* circuit is one which dissipates energy; consequently its subspace **S** is [,]-*negative*, or equivalently W is a *contraction*. A *lossless* circuit conserves energy, so **S** is *null*, and W is an *isometry*.

An aside is that some cross ratios actually occur in electrical engineering. For example, if subspaces $\mathbf{S}_S$ and $\mathbf{S}_L$ are associated with strictly passive circuits S and L, then

$$\mathbf{C}(\mathbf{S}_L, \mathbf{S}_S, \mathbf{S}'_L, \mathbf{S}'_S)$$

gives the power transfer ratio from S to L described by (2.4). References are [**HR**], [**H6**], [**BH7**].

A cascade connection as in Figure 3.1 is described mathematically by a linear mapping F acting on **K**. If M is energy conserving, then F is a [,]-isometry. Thus if C is described by subspace **S**, cascading by F produces a new subspace **S'** corresponding to the cascaded circuit.

Clearly we could Fourier transform this whole picture and obtain another L^2 space $\hat{\mathbf{K}} = \hat{\mathbf{K}}_+ + \hat{\mathbf{K}}_-$. Then the cascade mapping F on **K** Fourier transforms to a multiplication operator $\hat{F}$ on $\hat{\mathbf{K}}$. Since we are working with nothing more than multiplication by matrix-valued functions or vector-valued L^2, it looks as though we can compute everything pointwise and so face nothing but trivial mathematics problems.

Alas, this is not to be. The rub is that engineering makes a big issue of stable circuits and consequently of functions with no poles in the R.H.P. It turns out that an excellent way to capture this structure is with subspaces of L^2 which are invariant under the forward shifts. It turns out surprisingly that the classification and study of these subspaces has a remarkable amount of power. That is the subject of the next two chapters.

F. Homework. *Engineering problem.* A spectrum analyzer is calibrated by taking a certain number L of simple reference circiuts $S_1, S_2, S_3, \ldots, S_L$, where frequency response functions are computed exactly (from Maxwell's equations), and measuring them to obtain measured values $M_1, M_2, \ldots, M_L$. Typically $M_j \neq S_j$. We now take a circuit whose frequency response function S we wish to find; our analyzer measures M. What is S? How many reference measurements L are necessary to answer the question?

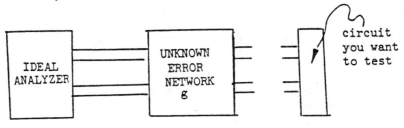

FIGURE 3.5. A two-port analyzer.

Math formulation. Model the problem, represented in Figure 3.5, as an ideal spectrum analyzer preceded by an unknown network which contains all of the errors. Our problem is to determine g, the F.R.F. of the error network; then the answer is $\mathbf{G}_g^{-1}(M) = S$.

The math problem (at each frequency) for an n-port analyzer is to uniquely determine a g in $GL(2n)$ which satisfies $\mathbf{G}_g(S_j) = M$, for all j.

Solution. To answer it make five "independent" measurements to get five reference spaces $\mathbf{S}_j$ and corresponding measured spaces $\dot{\mathbf{S}}_j$. Compute the cross ratios $\mathbf{C}(\mathbf{S}_1, \mathbf{S}_2, \mathbf{S}_3, \mathbf{S}_4)$, $\mathbf{C}(\dot{\mathbf{S}}_1, \dot{\mathbf{S}}_2, \dot{\mathbf{S}}_3, \dot{\mathbf{S}}_4)$, and the mapping $g_0 \colon \mathbf{S}_1 \leftrightarrow \dot{\mathbf{S}}_1$ which must intertwine them (since the model assumes that some g exists.) Use the construction behind Fact 3.D.2 to get one g which maps $\mathbf{S}_j \leftrightarrow \dot{\mathbf{S}}_j$. Another such $\tilde{g}$ satisfies $g_1^{-1}\tilde{g} \colon \mathbf{S}_j = \mathbf{S}_j$. Now the η in $GL(2n)$ which satisfy $\eta \mathbf{S}_j = \mathbf{S}_j$ have so limited a form that the additional restriction $\eta \mathbf{S}_5 = \mathbf{S}_5$ on a generically placed $\mathbf{S}_5$ determines η uniquely to within a scalar. This takes a brief calculation to check.

I hope the idea is clear from this. More details are in the article [**HSp**] with R. Speciale who designed TRW's two-port, very high frequency microwave spectrum analyzer. We gave algebra formulas and a description of five standard circuits which would calibrate your device. Present day calibration is done (in principle) with three standard (short circuit, a delay line, and a "through") and assumes that in the error network $g = \begin{pmatrix} \alpha & \beta \\ \kappa & \gamma \end{pmatrix}$, the entries $\beta = 0$ and $\kappa = 0$.

4. Representing Shift Invariant Subspaces

The classical Beurling-Lax theorem says that *any subspace* $\mathbf{M} \subset H_N^2$ *which is invariant under multiplication by* $e^{i\theta}$ *can be represented as*

$$(4.1) \qquad\qquad \mathbf{M} = gH_K^2$$

with a matrix-valued H^∞ *function whose values*

$$(4.2) \qquad\qquad g(e^{i\theta}) \text{ are isometries in } M_{NK} \text{ a.e.}$$

This g is uniquely determined up to a constant matrix. The theorem has an easy geometric proof due to Halmos and the surprising thing is that it has several powerful applications. In particular using the B-L theorem one can show the existence of Wiener-Hopf factorization, and one can prove the F. and M. Riesz theorem.

The main point of the next two lectures is that representation (4.1) for a given subspace $\mathbf{M}$ is highly nonunique unless we specify (4.2). The traditional restriction that $g(e^{i\theta})$ be an isometry is just *one way* of absorbing the freedom in the representation. In fact there are many different restrictions one could put on g besides this one. The remarkable discovery is that these lead to many other theorems in analysis.

This presentation begins with the most general representation. It puts forth our view that the Beurling-Lax-Halmos theorem is really a theorem about two subspaces $\mathbf{M}$ and $\mathbf{M}^\times$ of L_N^2, one invariant under $e^{i\theta}$ and one invariant under $e^{-i\theta}$. Henceforth we refer to such spaces as being *invariant* and *∗-invariant* respectively. Classical Beurling theory tells us that there are basically two types of invariant spaces, those which are

$$doubly\ invariant : \text{ invariant and ∗-invariant}$$
$$simply\ invariant : \bigcap_{n>0} e^{in\theta}\mathbf{M} = \{0\}.$$

In L_N^2 any invariant subspace is the direct sum of a doubly invariant and a simply invariant space. The prime example of a doubly invariant subspace of $L_N^2(\mathbf{T})$ is $L_N^2(E)$, the set of all functions supported on a closed subset E of the circle $\mathbf{T}$. In fact doubly invariant subspaces all have this form (cf. [**D**, Chapter 6]). The prime example of a simply invariant subspace of L_N^2 is H_N^2. In fact it is an easy exercise to check that *any invariant subspace of H_N^2 is simply invariant.*

35

An optimist would expect that *two disjoint simply invariant and ∗-simply invariant subspaces* $\mathbf{M}$ *and* $\mathbf{M}^\times$ *whose sum is* L_N^2 *can be represented as*

(4.3) $\mathbf{M} = gH_N^2, \qquad \mathbf{M}^\times = g\overline{H}_N^2$

for some $L_{N \times N}^\infty$ *function* g. *Moreover, if* $\mathbf{M}^\times = \mathbf{M}^\perp$ *one can take* g *to have unitary values.*

For most purposes in these talks you might as well assume that this is true. However, in general it is not true that g is uniformly bounded. Thus the representation we get is with a g in $L_{N \times N}^2$, and (4.3) becomes

(4.4) $\mathbf{M} = \operatorname{clos} gH_N^\infty, \qquad \mathbf{M}^\times = \operatorname{clos} g\overline{H}_N^\infty.$

Now we state a precise theorem. A bored reader might skip it since the basic idea has already been given. In fact *at first reading most readers should probably go from here to Chapter 5.*

This theorem allows for more general $\mathbf{M}$ and $\mathbf{M}^\times$ in that they do not have to be disjoint or fully span L_N^2. To handle these we define a *winding matrix* to be a function D in $L_{N \times N}^2$ of the form

$$D(e^{i\theta}) = \operatorname{diag}[e^{ik_1\theta}, e^{ik_2\theta}, \ldots, e^{ik_N\theta}]$$

with the integers $k_1 \leq k_2 \leq \cdots \leq k_N$ called *indices*. If $\mathbf{S}$ is a set we frequently denote its closure by $\mathbf{S}^-$, as well as $\operatorname{clos} \mathbf{S}$. Then a natural model for $\mathbf{M}$ and $\mathbf{M}^\times$ is DH_N^2 and $\overline{H}_N^2$. Another type of generality which causes no trouble is to replace L^2 with L^p and H^2 with H^p. In fact all theorems hold for weighted L^p spaces. Weights for these spaces are $N \times N$ matrix functions W which take *positive definite values* and which are *uniformly invertible*. Let Q denote the projection of L_N^p onto $\overline{H}_N^p$ along H_N^p. Such a space with norm

$$\int_0^{2\pi} (Wf, f)_{R^n}^{p/2} d\theta^{2/p} \triangleq \|f\|_W$$

will be denoted $L_N^p(W)$. As a set L_N^p and $L_N^p(W)$ are the same only the norms are different.

THEOREM 4.1. *Suppose* $\mathbf{M}$ *and* $\mathbf{M}^\times$ *are closed subspaces of* $L_N^p(W)$ *for* $1 < p < \infty$. *Then*

1. $\mathbf{M}$ *is simply invariant,* $\mathbf{M}^\times$ *is simply ∗-invariant,* $\dim(\mathbf{M} \cap \mathbf{M}^\times) < \infty$ *and* $\operatorname{codim}(\mathbf{M} + \mathbf{M}^\times) < \infty$ *if and only if*

2. *There is a function* $g \in L_{N \times N}^p$ *with* $g^{-1} \in L_{N \times N}^q$ *and a unique winding matrix* D *such that*

        (i) $\operatorname{clos}[gDH_N^\infty] = \mathbf{M}$,   (ii) $\operatorname{clos}[g\overline{H}_N^\infty] = \mathbf{M}^\times$,

        (iii) gQg^{-1} *is a bounded operator on* L_N^p.

The indices $[k_1, k_2, \ldots, k_N]$ *of the winding matrix* D *are determined by the formula*

(4.5) $\displaystyle\sum_{k_j < \alpha} (\alpha - k_j) = \dim[e^{i\alpha\theta}\mathbf{M}^\times \cap \mathbf{M}], \qquad \alpha = \ldots, -1, 0, 1, \ldots.$

If $\mathbf{M} + \mathbf{M}^\times = L_N^p$ *and* $\mathbf{M} \cap \mathbf{M}^\times = \{0\}$, *then g is determined up to a constant matrix.*

These general invariant subspace theorems are very closely related to the theory of Wiener-Hopf factorization, which has now been a tool in the theory of singular integral equations, Wiener-Hopf integral equations, and Toeplitz matrix equations for over half a century; a very recent survey is provided by the monograph of Clancey and Gohberg [**CG**]. Following [**CG**] we shall say that a matrix function A where both A and A^{-1} are in $L_{N \times N}^\infty$ has *a generalized right-standard factorization with respect to L_N^p* if

(i) The matrix function A can be factored in the form

$$(4.6) \qquad\qquad A = A_- D A_+$$

where $A_+ \in H_{N \times N}^q$, $A_+^{-1} \in H_{N \times N}^p$, $A_- \in \overline{H}_{N \times N}^p$, $A_-^{-1} \in \overline{H}_{N \times N}^q$, and $D(e^{i\theta}) = \mathrm{diag}[e^{ik_1\theta}, \ldots, e^{ik_N\theta}]$ is a winding matrix with indices $k_1 \leq k_2 \leq \cdots \leq k_N$ called the right-partial indices of A, and

(ii) The operator $A_- Q A_-^{-1}$ is bounded on L_N^p.

A similar definition of generalized left-factorization relative to L_N^p can be made; we, however, shall concentrate on right-factorization and say "generalized factorization" rather than "generalized right-factorization" in the sequel. The generalized factorization (4.6) will be called a *generalized canonical factorization* in case $k_1 = \cdots = k_N = 0$. If A is a function in $L_{N \times N}^\infty$, we let T_A denote the Toeplitz operator

$$T_A \colon f \to P(Af), \qquad f \in H_N^p,$$

where $P \colon L_N^p \to H_N^p$ is the (bounded) projection onto H_N^p along $\overline{H}_N^p$. The following factorization results are stated and proved in the monograph [**CG**] (in terms of singular integral operators rather than Toeplitz operators), where they are attributed to H. Widom and Simonenko. Our main point is that they are immediate corollaries of our invariant subspace theorem (4.1). Let $\mathrm{GL}_{N \times N}^\infty$ denote those functions A with both A and A^{-1} in $L_{N \times N}^\infty$.

COROLLARY 4.2 (GENERALIZED CANONICAL WIENER-HOPF FACTORIZATION). *The function A in $\mathrm{GL}_{N \times N}^\infty$ has a generalized canonical factorization with respect to L_N^p if and only if T_A is invertible on H_N^p.*

COROLLARY 4.3 (GENERALIZED WIENER-HOPF FACTORIZATION). *The function A in $\mathrm{GL}_{N \times N}^\infty$ has a generalized factorization $A = A_- D A_+$ with respect to L_N^p if and only if the Toeplitz operator T_A is Fredholm as an operator on H_N^p. In this case the indices $\{k_1, k_2, \ldots, k_N\}$ of the winding matrix D are uniquely determined by the formula*

$$\sum_{k_j \leq \alpha} (\alpha - k_j) = \dim(e^{i\alpha\theta} \overline{H}_N^p \cap A H_N^p), \qquad \alpha = \ldots, -1, 0, 1, \ldots.$$

Moreover, these Wiener-Hopf factorization theorems in effect are actually equivalent to the Invariant Subspace Theorem 4.1 in the sense that one can easily derive Theorem 4.1 from Corollary 4.3. By this remark, one sees that in general one cannot take the representing function g to be bounded with bounded inverse, since it is known (see [**CG**, p. 141]) that in general one cannot take the factors $A_\pm^{\pm 1}$ in Corollaries 4.2 and 4.3 to be bounded.

We now sketch proofs. Since the main idea is revealed in the canonical case we take $k_1 = k_2 = \cdots = k_N = 0$.

PROOF OF THEOREM 4.1. The proof that $(2) \Rightarrow (1)$ is straightforward. Indeed, suppose $g \in L^p_{N \times N}$ and gQg^{-1} a bounded operator on L^p_N. Set $\mathbf{M} = [gH^\infty_N]^-$ and $\mathbf{M}^\times = [g\overline{H}^\infty_N]^-$ where the closures are in L^p_N.

If $g \in \mathrm{GL}^\infty_{N \times N}$, it is clear that $\mathbf{M}$ is simply invariant and that $\mathbf{M}^\times$ is simply *-invariant. For the general case one needs a slightly technical argument.[1] Next note that functions of the form gp, where $p(e^{i\theta}) = \sum_{k=-R}^{R} e^{ik\theta} X_k$ is a $\mathbb{C}^N$-valued trigonometric polynomial, are dense in L^p_N, and

$$gQg^{-1}\left(g\left\{\sum_{-R}^{R} e^{ik\theta} X_k\right\}\right) = g\left\{\sum_{-R}^{-1} e^{ik\theta} X_k\right\} \in \mathbf{M}^\times$$

while

$$[I - gQg^{-1}]\left(g\left\{\sum_{-R}^{R} e^{ik\theta} X_k\right\}\right) = g\left\{\sum_{0}^{R} e^{ik\theta} X_k\right\} \in \mathbf{M}.$$

The fact that gQg^{-1} extends to a bounded operator on L^p_N means that there is a bounded projection of L^p_N onto $\mathbf{M}^\times$ along $\mathbf{M}$, so $L^p_N = \mathbf{M}^\times \dotplus \mathbf{M}$. This proves (1). Here $\dotplus$ denotes the direct sum of two spaces.

Conversely, suppose $\mathbf{M}$ and $\mathbf{M}^\times$ are two subspaces of L^p_N which satisfy (1) in Theorem 4.1. We shall show that the following recipe constructs the desired function g: Let $\mathbf{L}$ be the subspace $e^{i\theta}\mathbf{M}^\times \cap \mathbf{M}$. We shall show that $\dim \mathbf{L} = N$, so $\mathbf{L}$ has a basis of N functions $\{y_1, \ldots, y_N\}$. Let g be the $N \times N$ matrix function with jth column equal to y_j. Then g is the desired "Beurling-Lax-Halmos" representer.

By hypothesis $L^p_N = \mathbf{M}^\times \dotplus \mathbf{M}$, so also

$$L^p_N = e^{i\theta}\mathbf{M}^\times \dotplus e^{i\theta}\mathbf{M}.$$

Since $e^{i\theta}\mathbf{M} \subset \mathbf{M}$, we conclude from this that

$$\mathbf{M} = L^p_N \cap \mathbf{M} = (e^{i\theta}\mathbf{M}^\times \cap \mathbf{M}) \dotplus e^{i\theta}\mathbf{M} = \mathbf{L} \dotplus e^{i\theta}\mathbf{M},$$

---

[1]Choose a bounded scalar outer function s so that sg is bounded. Then

$$\log(|s|^{-2}\|(g^*g)^{-1}\|)$$

is integrable since $g^{-1} \in L^q_{N \times N}$; so, by Devinatz's criterion (see |**Hls**|), $|s|^2 g^* g$ has an outer factorization $|s|^2 g^* g = G^* G$ where $G \in H^\infty_{N \times N}$ is outer. Then $sgf \to Gf$ is an L^2_N-isometry from $[gsH^2_N]^-$ onto $[GH^2_N]^- = H^2_N$ which intertwines the respective shift operators; since H^2_N is simply invariant for multiplication by $e^{i\theta}$, so is $[gsH^2_N]^- \subset L^2_N$. From this it follows that $\mathbf{M} = [gH^\infty_N]^- \subset L^p_N$ is simply invariant. The simple *-invariance of $\mathbf{M}^\times$ follows similarly.

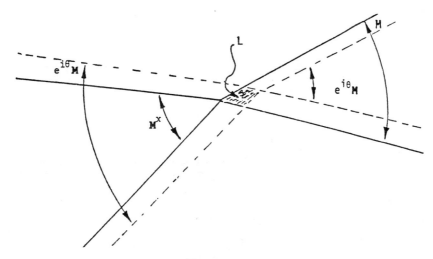

FIGURE 4.1.

where we have set $L = e^{i\theta}M^{\times} \cap M$. A picture helps me; I don't know if it helps anyone else (see Figure 4.1). If we iterate this identity, we get

$$\mathbf{M} = \mathbf{L} \dotplus e^{i\theta}(\mathbf{L} \dotplus e^{i\theta}\mathbf{M}) = \mathbf{L} \dotplus e^{i\theta}\mathbf{L} \dotplus e^{2i\theta}\mathbf{M}.$$

This plus the similar computation for $\mathbf{M}^{\times}$ gives

$$(4.7) \qquad \begin{aligned} \mathbf{M} &= \mathbf{L} \dotplus e^{i\theta}\mathbf{L} \dotplus \cdots \dotplus e^{i\alpha\theta}\mathbf{L} \dotplus e^{i[\alpha+1]\theta}\mathbf{M}, \\ \mathbf{M}^{\times} &= e^{-i\theta}\mathbf{L} \dotplus e^{-2i\theta}\mathbf{L} \dotplus \cdots \dotplus e^{-i\beta\theta}\mathbf{L} \dotplus e^{-i\beta\theta}\mathbf{M}^{\times} \end{aligned}$$

for all α and $\beta \geq 0$.

From here on the formal argument is very intuitive. Simple invariance says that the remainder terms in these expressions go to zero. With some work[2] one

---

[2]Expressions (4.7) imply

$$(4.7') \qquad L_N^p = \mathbf{M}^{\times} \dotplus \mathbf{M} = e^{-\beta\theta}\mathbf{M}^{\times} \dotplus \left[\dotplus_{k=-\beta}^{\alpha} e^{ik\theta}\mathbf{L}\right] \dotplus e^{i(\alpha+1)\theta}\mathbf{M}$$

for all $\beta, \alpha \geq 0$. In particular there is a bounded projection of L_N^p onto $\mathbf{L}$ along $\mathbf{M}^{\times} \dotplus e^{i\theta}\mathbf{M}$ which we denote by $P_{\mathbf{L}}$, and a bounded projection onto $\mathbf{M}$ along $\mathbf{M}^{\times}$ which we denote by $P_{\mathbf{M}}$. A straightforward computation using (4.7) gives that the projection onto $e^{ik\theta}\mathbf{L}$ along $e^{ik\theta}\mathbf{M}^{\times} \dotplus e^{i(k+1)\theta}\mathbf{M}$ is $e^{ik\theta}P_{\mathbf{L}}e^{ik\theta}$, and that the projection onto $e^{i(\alpha+1)\theta}\mathbf{M}$ along $\mathbf{M}^{\times} \dotplus [\dotplus_{k=0}^{\alpha} e^{ik\theta}\mathbf{L}]$ is given by $P_\alpha \equiv e^{i(\alpha+1)\theta}P_{\mathbf{M}}e^{-i(\alpha+1)\theta}$. In particular, we see that $\|P_\alpha\| \leq \|P_\beta\| < \infty$ for all $\alpha \geq 0$. It follows from an argument in [CFS] that $[P_\alpha]$ converges strongly to a bounded projection R, where $\operatorname{Ran} R = \bigcap_{\alpha \geq 0} \operatorname{Ran} P_\alpha$ and $\operatorname{Ker} R = \{\bigcup_{\alpha \geq 0} \operatorname{Ker} P_\alpha\}^{-}$. Since $\mathbf{M}$ is simply invariant, $\bigcap_{\alpha \geq 0} \operatorname{Ran} P_\alpha = \bigcap_{\alpha \geq 0} e^{i(\alpha+1)\theta}\mathbf{M} = \{0\}$, so $\{P_\alpha\}$ converges strongly to 0 as $\alpha \to \infty$. Similarly, $\{P_\beta^x\}$ converges strongly to 0 as $\beta \to \infty$ where $P_\beta^x \equiv e^{i\beta\theta}(1-P_\beta)e^{i\beta\theta}$ is the projection onto $e^{-i\beta\theta}\mathbf{M}^{\times}$ along $[\dotplus_{k=-\beta}^{-1} e^{ik\theta}\mathbf{L}] \dotplus \mathbf{M}$. We conclude that

$$(4.8) \qquad f = \lim_{\substack{\beta \to \infty \\ \alpha \to \infty}} \sum_{k=-\beta}^{\alpha} e^{ik\theta}P_{\mathbf{L}}(e^{-ik\theta}f)$$

for any $f \in L_N^p$ where the infinite direct sum is in the sense of a *Schauder decomposition*.

can actually prove that

$$(4.9) \qquad L_N^p = \overset{\infty}{\underset{-\infty}{\dotplus}} \, e^{ik\theta}\mathbf{L}$$

and

$$(4.9') \qquad \mathbf{M} = \overset{\infty}{\underset{k=0}{\dotplus}} \, e^{ik\theta}\mathbf{L}, \qquad \mathbf{M}^\times = \overset{k=1}{\underset{-\infty}{\dotplus}} \, e^{ik\theta}\mathbf{L}$$

where the infinite direct sum is in the sense of a Schauder decomposition.

Now it is time to look at an example. Let $\mathbf{M} = H_k^p$ and $\mathbf{M}^\times = \overline{H}_k^p$. Then $\mathbf{L} = e^{i\theta}\overline{H}_k^p \cap H_k^p = $ constant functions in $L_k^p \triangleq \tilde{\mathbf{L}}$. Formulas (4.9) say

$$(4.10) \qquad L_k^p = \cdots + \mathrm{CONST}\, e^{-i\theta} + \mathrm{CONST} + \mathrm{CONST}\, e^{i\theta} + \cdots$$

which essentially says that every f in L_k^p has Fourier decomposition.

To finish the proof all that we must do is put the "reference" decomposition (4.10) into correspondence with the decomposition (4.9) produced by the subspace $\mathbf{M}$ we are studying. This is easy. Let $\tilde{\mathbf{L}}$ be a Banach space isomorphic to $\mathbf{L}$, and let $\tilde{g}\colon \tilde{\mathbf{L}} \to \mathbf{L}$ be an isomorphism. (We shall see in the end that we may take $\tilde{\mathbf{L}} = \mathbb{C}^N$.) One can think of $\tilde{g}$ as putting $\tilde{\mathbf{L}}$ in (4.10) in correspondence with $\mathbf{L}$ in (4.9); from this correspondence the rigidity of formulas (4.9) and (4.10) induces a mapping from $L_{\tilde{\mathbf{L}}}^p$ to L_N^2 by a function g. Indeed g is defined by

$$g(e^{i\theta})x = (\tilde{g}x)(e^{i\theta}), \qquad x \in \tilde{\mathbf{L}},$$

so $g(e^{i\theta})$ is defined a.e. as an operator from $\tilde{\mathbf{L}}$ into $\mathbb{C}^N$. By the closed graph theorem, multiplication by the operator function g is bounded as an operator from $\tilde{\mathbf{L}}$ to L_N^p, so

$$(4.11) \qquad \frac{1}{2\pi}\int_0^{2\pi} \|g(e^{i\theta})\|_{\mathbf{L} \times \mathbb{C}^N}^p \, d\theta < \infty.$$

Let $L_{\tilde{\mathbf{L}}}^\infty$ denote the Banach space of weak-$*$ measurable uniformly bounded $\tilde{\mathbf{L}}$-valued functions, $\overline{H}_{\tilde{\mathbf{L}}}^\infty$ the subspace of boundary values of functions analytic on the exterior of the disk and vanishing at ∞. Thus

$$H_{\tilde{\mathbf{L}}}^\infty = \text{weak-}* \text{ closure of the span of } \{e^{ik\theta}x | x \in \tilde{\mathbf{L}}, \; k = 0, 1, 2, \ldots\}$$

and

$$\overline{H}_{\tilde{\mathbf{L}}}^\infty = \text{weak-}* \text{ closure of the span of } \{e^{ik\theta}x | x \in \tilde{\mathbf{L}}, \; k = -1, -2, \ldots\}.$$

We see from (4.11) that multiplication by g maps $L_{\tilde{\mathbf{L}}}^\infty$ into L_N^p. It follows immediately from the decompositions (4.10) and the definition of g that

$$(4.12a) \qquad [gL_{\tilde{\mathbf{L}}}^\infty]^- = L_N^P,$$

$$(4.12b) \qquad [gH_{\tilde{\mathbf{L}}}^\infty]^- = \mathbf{M},$$

$$(4.12c) \qquad [g\overline{H}_{\tilde{\mathbf{L}}}^\infty]^- = \mathbf{M}^\times.$$

Our next goal is to show that $\dim \tilde{\mathbf{L}} = N$, so then we may take $\tilde{\mathbf{L}} = \mathbb{C}^N$. If $\dim \tilde{\mathbf{L}} < N$ on a set of positive measure, one could construct a nonzero function in L_N^p which is not in $[gL_{\tilde{\mathbf{L}}}^\infty]^-$, in violation of (4.12a). In a similar way, to show that $\dim \tilde{\mathbf{L}} \geq N$, it suffices to show that $\ker g(e^{i\theta}) = (0)$ for a.e. θ, or that multiplication by g is injective as an operator from $L_{\tilde{\mathbf{L}}}^\infty$ into L_N^p. So suppose that $gf = 0$ for some f in $L_{\tilde{\mathbf{L}}}^\infty$. Since $\tilde{g}$ is invertible from $\tilde{\mathbf{L}}$ to $\mathbf{L}$, the map g restricted to finite sums in the decomposition (4.10) has range equal to the same finite sums in (4.9) and such restrictions are invertible (since $\tilde{g}: \tilde{\mathbf{L}}$ to $\mathbf{L}$ is invertible). This and a limit argument will imply that any f with $gf = 0$ has zero Fourier coefficients; so $f = 0$.

Now we can identify $\tilde{\mathbf{L}}$ with $\mathbb{C}^N$ and we can view g as an $N \times N$ matrix-valued function. From (4.11) we see that $g \in L_{N \times N}^p$, and from (4.12b) and (4.12c) we see that $\mathbf{M} = [gH_N^\infty]^-$ and $\mathbf{M}^\times = [g\overline{H}_N^\infty]^-$. It remains to show that $g^{-1} \in L_{N \times N}^q$.

To see this, we resort to a dual of the above analysis. The spaces L_N^p and L_N^q are duals of each other, where the duality pairing is given by the bilinear form

$$\langle f, g \rangle = \frac{1}{2\pi} \int_0^{2\pi} g(e^{i\theta})^T f(e^{i\theta}) \, d\theta.$$

The decomposition $L_N^p = \mathbf{M}^\times \dotplus \mathbf{M}$ is equivalent to a dual decomposition $L_N^q = \mathbf{M}^{\times \perp} \dotplus \mathbf{M}^\perp$ for L_N^q. Here $\mathbf{M}^\perp$ is the annihilator of $\mathbf{M}$ in L_N^q with respect to the above pairing. It is straightforward to check that $\mathbf{M}$ being simply invariant and $\mathbf{M}^\times$ being simply $*$-invariant is equivalent to $\mathbf{M}^\perp$ being simply invariant and $\mathbf{M}^{\times \perp}$ being simply $*$-invariant. Thus by the part of the theorem already proved, we know that there is an $h \in L_{N \times N}^q$ such that

$$\mathbf{M}^\perp = [hH_N^\infty]^- \quad \text{and} \quad \mathbf{M}^{\times \perp} = [h\overline{H}_N^\infty]^-$$

(the closures in L_N^q). It follows next that

$$\langle h^T g u, v \rangle = 0 \quad \text{for all } u, v \text{ in } H_N^\infty$$

and

$$\langle h^T g u, v \rangle = 0 \quad \text{for all } u, v \text{ in } \overline{H}_N^\infty,$$

where $\langle \, , \, \rangle$ is the $L_N^p \Leftrightarrow L_N^q$ pairing (in this case $p = 1$ and $q = \infty$) mentioned above. In particular all Fourier coefficients except the coefficient of $e^{+i\theta}$ for the L^1-function $h^T g$ vanish, so $[h^T g](e^{i\theta}) = e^{+i\theta} c$ for some $c \in \mathbf{M}_N$. On the other hand, we know that $[h^T g](e^{i\theta})$ must be invertible a.e., so $c \in \mathrm{GL}(N)$. Thus $g^{-1}(e^{i\theta}) = c^{-i\theta} c^{-1} h^T(e^{i\theta}) \in L_{N \times N}^q$ as claimed. Also that $gQg^{-1}: L_N^\infty \to L_N^1$ actually extends to define a bounded operator on L_N^p follows easily from the boundedness of the projection $P_{\mathbf{M}^\times}$ onto $\mathbf{M}^\times$ along $\mathbf{M}$.

Finally, suppose g_1 is another representer satisfying conditions 2(i)–2(iii) in Theorem 4.1. Then, by the computation above $\mathbf{M} = [gH_N^\infty]^- \subset L_N^p$ and $\mathbf{M}^\times = [e^{i\theta} g_1^{-1^T} H_N^\infty] \subset L_N^q$, while $\mathbf{M}^\times = [g\overline{H}_N^\infty]^-$ and $\mathbf{M}^{\times \perp} = [e^{i\theta} g_1^{-1^T} \overline{H}_N^\infty]^-$. By the argument given above, we see that $e^{+i\theta} g_1^{-1} g = e^{+i\theta} c$ for some constant $c \in$

$GL(N)$, so $g(e^{i\theta}) = g_1(e^{i\theta})c$. This proves the uniqueness assertion in Theorem 4.1 and completes the proof.

We now derive the canonical case of generalized Wiener-Hopf factorization as a direct consequence of Theorem 4.1.

PROOF OF COROLLARY 4.2. We are given a function A in $GL_{N\times N}^{\infty}$. It is a direct check to see that the Toeplitz operator T_A is invertible on L_N^p if and only if L_N^p has the direct sum decomposition

$$L_N^p = \overline{H}_N^p \dotplus A H_N^p.$$

Clearly $\overline{H}_N^p$ is simply $*$-invariant and AH_N^p is simply invariant. By Theorem 4.1 there is a function $g \in L_{N\times N}^p$ with $g^{-1} \in L_{N\times N}^q$ such that $\overline{H}_N^p = [g\overline{H}_N^\infty]^-$ and $AH_N^p = [gH_N^\infty]^-$ and gQg^{-1} is a bounded operator on L_N^p. Set $A_- = g$ and $A_+ = g^{-1}A$. Then it is a direct check that $A = A_-A_+$ is the desired Wiener-Hopf factorization.

It is also possible to derive the invariant subspace Theorem 4.1 easily from its Corollary 4.2, with the help of the classical Beurling-Lax theorem, as follows. Thus the application of Theorem 4.1 to factorization is somewhat close to the theorem itself.

PROOF OF THEOREM 4.1 (USING COROLLARY 4.2). We are given a simply $*$-invariant subspace $\mathbf{M}^\times$ such that $L_N^p = \mathbf{M}^\times \dotplus \mathbf{M}$. By the classical Beurling-Lax theorem (see [Hls] for a discussion of the $p \neq 2$ case), we can write $\mathbf{M}^\times = \Psi \overline{H}_N^p$ and $\mathbf{M} = \Gamma H_N^p$ for unitary valued functions Ψ and Γ in $L_{N\times N}^\infty$. Then we have another convenient decomposition for L_N^p,

$$L_N^p = \Psi^{-1} L_N^p = \overline{H}_N^p \dotplus \Psi^{-1} \Gamma H_N^p.$$

It follows that $T_{\Psi^{-1}\Gamma}$ is invertible, so by the corollary, $\Psi^{-1}\Gamma$ has a canonical Wiener-Hopf factorization $\Psi^{-1}\Gamma = A_-A_+$. It is a direct check that $g \triangleq \Psi A_-$ satisfies the conditions of Theorem 4.1 as needed.

5. Applications to Factorization, Interpolation, and Approximation

In the preceding chapter we gave a general representation theorem for invariant subspaces of L_N^2. In this chapter we describe the many analysis theorems which follow from it and from a few basic facts about projective space. A list of results which (after preliminaries) actually constitutes an outline for this lecture is

A. Interpolation.
B. Toeplitz Corona Theorem.
C. Factorization.
D. Disk problems.
E. More on interpolation.

The form of the representation theorem which is the workhorse of this chapter is set in L_N^2 endowed with a sesquilinear form $[\ ,\ ]$ gotten from a sesquilinear form $[\ ,\ ]_{\mathbb{C}^N}$ on $\mathbb{C}^N$ by

$$[f, h] = \frac{1}{2\pi} \int_0^{2\pi} [f(e^{i\theta}), h(e^{i\theta})]_{\mathbb{C}^N}\, d\theta.$$

We assume that $[\ ,\ ]_{\mathbb{C}^N}$ is nondegenerate with $k_+ = \mathbb{C}^m$ and $k_- = \mathbb{C}^n$ in which case L_N^2, $[\ ,\ ]$ is a Krein space with a decomposition $\mathbf{K}_+ = L_m^2$ and $\mathbf{K}_- = L_n^2$. Here $m + n = N$. An immediate corollary of Theorem 4.1 is:

THEOREM 5.1. *If* $\mathbf{M}^\times$ *is a closed simply invariant subspace of* L_N^2; *if* $\mathbf{M}'$ *does not intersect* $\mathbf{M}$; *if* $\mathbf{M} + \mathbf{M}' = L_N^2$, *then there is a* g *in* $L_{N \times K}^2$ *which represents* $\mathbf{M}$ *as*

$$\mathbf{M} = \operatorname{clos} g H_K^\infty$$

and whose boundary values $g(e^{i\theta})$ *are* $[\ ,\ ]_{\mathbb{C}^K}$ *to* $[\ ,\ ]_{\mathbb{C}^N}$ *isometries a.e. Always* $K \leq N$; *if* $\mathbf{M}$ *is full range*[1] $K = N$.

PROOF. Our pedagogy cheats in that this theorem does not exactly follow from Theorem 4.1. The added generality here is that we are not now assuming that $\mathbf{M}'$ is simply $*$-invariant. Thus to prove this theorem one should actually go

---

[1]Full range is a nondegeneracy condition on $\mathbf{M}$ which says that the vector space

$$V_{z_0} \triangleq \{f(z_0) : f \in \mathbf{M}\}$$

has dimension N for some z_0 contained in the unit disk. One can show that if one z_0 has this property all but a set of isolated z_0 have this property.

through the proof of Theorem 4.1 again. Namely, take $\mathbf{L} = (e^{i\theta}\mathbf{M'}) \cap \mathbf{M}$, repeat the Theorem 4.1 argument, and ultimately establish that $\mathbf{L}$ has dimension K.

We now demonstrate that Theorem 5.1 follows directly from Theorem 4.1 if $\mathbf{M'}$ is *-simply invariant (which incidentally is equivalent to $\mathbf{M}$ having full range). The proof goes by representing $\mathbf{M}$ and $\mathbf{M'}$ with g and observing that

$$\text{clos } gH_N^\infty = \mathbf{M} = \mathbf{M''} = (J\mathbf{M'})^\perp = (JgH_N^{\infty\perp})^\perp.$$

Here $[\ ,\ ]_{\mathbb{C}^N} = \langle J, \ \rangle_{\mathbb{C}^N}$ with $J^2 = I$ and $J^* = J$. Now $x \in (JgH_N^{\infty\perp})^\perp$ means

$$0 = (x, JgH_N^{\infty\perp}) = (g^*Jx, H_N^{\infty\perp}),$$

which means $g^*Jx \in H_N^2$. Thus we have

$$\text{clos } g^*JgH_N^\infty = H_N^2,$$

so $g^*Jg = C$, a constant matrix. The Cholesky decomposition $C = L^*JL$ yields $L^{*-1}g^*JgL = J$ which means that gL is a J-isometry.   Q.E.D.

COROLLARY 5.2 (BEURLING-LAX). *If* $[\ ,\ ]$ *is the standard* L_N^2 *inner product, then any closed simply invariant full range* $\mathbf{M}$ *has a representation* $\mathbf{M} = gH_N^2$ *with* $g \in L_{N \times N}^\infty$ *and* $g(e^{i\theta})$ *a unitary matrix a.e.*

PROOF. For the usual positive definite bilinear form $[\ ,\ ]$ the orthogonal complement $\mathbf{M'}$ of $\mathbf{M}$ is always disjoint from it and $\mathbf{M} + \mathbf{M'} = L_N^2$.

Now we turn to applications of invariant subspace representations. Before moving to great generality we give the basic idea behind several subjects by treating the simplest special examples.

A. Interpolation. *Given* $z_j, w_j \in \mathbb{C}$ *with modulus less than one, compute* $\mathbf{T}^l \triangleq \{f \in \overline{\mathbf{B}}H^{\infty;l} : f(z_j) = w_j\}$. *Is* $\mathbf{T}^l$ *empty? Here* $j = 1, \ldots, p$.

Solution. Set $[x, y]_{\mathbb{C}^2} = x_1\bar{y}_1 - x_2\bar{y}_2$ and as usual let $[\ ,\ ]$ denote the form induced by it on L_2^2. Set

$$\mathbf{M} = \left\{ \begin{pmatrix} u \\ v \end{pmatrix} \in H_2^2 : u(z_j) = w_j v(z_j) \right\} \subset H_2^2.$$

If $\mathbf{M} \cap \mathbf{M'} = \{0\}$ by Theorem 5.1 we can represent $\mathbf{M}$ as $\text{clos } gH_2^\infty$ where $g(e^{i\theta})$ is a $[\ ,\ ]_{\mathbb{C}^2}$ isometry a.e. in θ. In other words each $g(e^{i\theta})$ belongs to the Lie group $U(1,1)$. Since the codimension of $\mathbf{M}$ in H_2^2 is finite (it equals p), g is a rational function; however, this fact plays no critical role in the argument. In particular, g and $g^{-1} \in L^\infty$; so multiplication by g is bounded and invertible as an operator on L_2^2.

The representer g for $\mathbf{M}$ gives a complete solution to the interpolation problem:

THEOREM 5.A.1. *Suppose* $\mathbf{M} \cap \mathbf{M'} = \{0\}$. *Set* $l = $ *the negative signature of* $\mathbf{M'} \cap H_2^2$. *Then* $\text{clos } \mathbf{T}^l = \mathbf{G}_g(\overline{\mathbf{B}}H^\infty)$. *When* $\tilde{l} < l$ *the set* $\mathbf{T}^{\tilde{l}}$ *is empty. Here* $\text{clos } \mathbf{T}^l$ *refers to the closure of* $\mathbf{T}^l$ *as a subset of* L^∞.

Later we shall show various ways to compute f. Now we prove the theorem.

PROOF. The proof amounts to checking three things, most of which are merely conversions from angle operator to subspace terms. Each one takes only a few lines to check; so, the proof is indeed extremely simple.

(i)   clos $\mathbf{T}^l$ is the set of all angle operators for invariant negative subspaces of $\mathbf{M}$ which have negative cosignature l in H_2^2.

This follows immediately from Fact 3.C.4 and the definition of $\mathbf{M}$. After all a subspace $\mathbf{S}$ of this type has an angle operator F which is multiplication by an f in $\overline{\mathbf{B}}H^{\infty;l}$. Let D denote the domain of the angle operator. The fact that $\mathbf{S}$ is contained in $\mathbf{M}$ says precisely that for any v in D the function $\binom{fv}{v}$ satisfies $f(z_j)v(z_j) = w_j v(z_j)$. In other words $f(z_j) = w_j$. The only exception is if z_j is a pole of f, but this rare case is subsumed in the act of closing $\mathbf{T}^l$.

(ii)   The invariant maximal negative subspaces of $\mathbf{M} = g($the invariant maximal negative subspaces of $H_2^2)$.

Since g is a multiplication operator it maps invariant subspaces to invariant subspaces. Since g is a [,] isometry it maps maximal negative spaces to maximal negative spaces.

(iii)   Each invariant maximal negative subspace of $\mathbf{M}$ has negative cosignature l in H_2^2, where l = the negative signature of $\mathbf{M}' \cap H_2^2$.

This is Fact 3.C.5. Combine (i), (ii), and (iii) to get

        clos $\mathbf{T}^l$ is the set of all angle operators for g (the invariant maximal negative subspaces of H_2^2).

Since (by Fact 3.C.4) the angle operators for the invariant maximal negative subspaces of H_2^2 correspond precisely to the unit ball $\overline{\mathbf{B}}H^\infty$ of H^∞, Fact 3.C.1 implies that the f in $\mathbf{T}^l$ all have the form

$$f(e^{i\theta}) = [\alpha(e^{i\theta})b(e^{i\theta}) + \beta(^{i\theta})][\kappa(e^{i\theta})b(e^{i\theta}) + \gamma(e^{i\theta})]^{-1}$$

where $g = \binom{\alpha\ \beta}{\kappa\ \gamma}$. Here b sweeps through $\overline{\mathbf{B}}H^\infty$. This proves Theorem 5.A.1.

The most standard way to compute the integer l in Theorem 5.A.1 is using the "Pick" matrix $\Lambda_{\{z,w\}}$ defined by

$$\Lambda_{\{z,w\}} = \left\{ \frac{1 - w_i \bar{w}_j}{1 - z_i \bar{z}_j} \right\}^p_{i,j=1}.$$

In this context z and w denote the tuple of points $(z_1, z_2, \ldots, z_p)$ and $(w_1, w_2, \ldots, w_p)$. The formula for l is given by

THEOREM 5.A.2. The space $\mathbf{M} \cap \mathbf{M}'$ is not 0 if and only if $\Lambda_{\{z,w\}}$ has a null vector. The negative signature of $\mathbf{M}' \cap H_2^2$ equals the number of negative eigenvalues of $\Lambda_{\{z,w\}}$.

PROOF. The thing which makes the proof easy is that one can write down a basis for $\mathbf{M}' \cap H_2^2$. The basis uses the Szegö reproducing kernel $\kappa_\xi(e^{i\theta})$ for H^2, a function defined by

$$(5.1) \qquad \kappa_\xi(e^{i\theta}) = \frac{1}{2\pi} \frac{1}{1 - \bar{\xi}e^{i\theta}}$$

for each $\xi \in \mathbb{C}$ with $|\xi| < 1$. Certainly κ_ξ is in H^2 and it has the reproducing property, namely

$$f(\xi) = (f, \kappa_\xi)_{L^2}$$

for each $f \in H^2$. This follows from the Cauchy integral theorem via

$$(f, \kappa_\xi) = \frac{1}{2\pi} \int_0^{2\pi} f(e^{i\theta}) \frac{1}{1 - \xi e^{-i\theta}} \, d\theta = \frac{1}{2\pi i} \int_T f(z) \frac{1}{z - \xi} \, dz = f(\xi).$$

We claim that

$$b_j(e^{i\theta}) = \frac{1}{1 - \bar{z}_j e^{i\theta}} \begin{pmatrix} 1 \\ \bar{w}_j \end{pmatrix}$$

is a basis for $\mathbf{M}'$. To check this merely compute

$$(5.2) \qquad \left[\begin{pmatrix} u \\ v \end{pmatrix}, b_j \right] = (u, \kappa_{z_j})_{L^2} - w_j(v, \kappa_{z_j})_{L^2} = u(z_j) - w_j v(z_j) = 0$$

for all $\begin{pmatrix} u \\ v \end{pmatrix}$ in $\mathbf{M}$ to see that each b_j is in $\mathbf{M}'$. This same formula says that any $\begin{pmatrix} u \\ v \end{pmatrix}$ in H_2^2 which is $[\, , \,]$ orthogonal to all of the b_j's is in $\mathbf{M}$. Consequently (5.2) says span $\{b_1, \ldots, b_p\} \subset \mathbf{M}'$ and span $\{b_1, \ldots, b_p\}' \cap H_2^2 \subset \mathbf{M}$. The first containment implies span $\{b_1, \ldots, b_p\}' \supset \mathbf{M}'' = \mathbf{M}$, so span $\{b_1, \ldots, b_p\}' \cap H_2^2 = \mathbf{M}$ which says that the b_j are a basis for $\mathbf{M}' \cap H_2^2$.

Once we have a basis for $\mathbf{M}' \cap H_2^2$, computing its $[\, , \,]$ negative signature is easy. First we compute the Gramian $[b_i, b_j]$ for the basis b_j. Its entries are

$$[b_i, b_j] = \left[\begin{pmatrix} 1 \\ \bar{w}_i \end{pmatrix}, \begin{pmatrix} 1 \\ \bar{w}_j \end{pmatrix} \right]_{\mathbb{C}^2} (\kappa_{z_i}, \kappa_{z_j})$$

$$= (1 - \bar{w}_i w_j) \frac{1}{1 - \bar{z}_i z_j}$$

from which we see that the Gramian equals the "Pick matrix" $\Lambda_{\{z,w\}}$!

For any vectors x, y in $\mathbf{M}' \cap H_2^2$ we expand them in the basis b_j and see that

$$(5.3) \qquad [x, y] = \sum_{i,j=1}^p x_j[b_i, b_j]\bar{y}_j = \left(\Lambda \begin{pmatrix} x_1 \\ x_p \end{pmatrix}, \begin{pmatrix} y_1 \\ y_p \end{pmatrix} \right)_{\mathbb{C}^p}.$$

Consequently, there is a nontrivial x in $\mathbf{M} \cap \mathbf{M}'$ if and only if $[x, y] = 0$ for all $y \in \mathbf{M}' \cap H_2^2$, if and only if $\Lambda x = 0$. This proves the first part of the theorem. By (5.3) the negative signature of $[\, , \,]$ on $\mathbf{M}' \cap H_2^2$ equals the negative signature of the bilinear form $(\Lambda,)_{\mathbb{C}^p}$ on $\mathbb{C}^p$. This, of course, equals the number of negative eigenvalues of $\Lambda_{\{z,w\}}$.

This completes a study of the case of interpolation theory via a method which generalizes enormously. A reader might wonder about a few further detailed points such as

(1) What happens when $\mathbf{M} \cap \mathbf{M}' \neq \{0\}$?

(2) Can one get explicit formulas for g?

We shall now sketch the solution to (1) since it is much simpler than what we have just been through. Item 2 we leave as Exercise 5.1 (solution in Chapter 5.E). The article by Francis-Zames and myself carries out such computations in substantial detail at greater generality. Then Doyle [**D**] did a "state space" version and used it in the earliest version of his control system design program at Honeywell. Later it was replaced by Glover's "state space" formulas.

Suppose $\mathbf{M} \cap \mathbf{M}' \neq \{0\}$. Consider the subspace

$$\mathbf{N} = \bigvee_{n \geq 0} e^{in\theta} (\mathbf{M} \cap \mathbf{M}')$$

of $\mathbf{M}$. It is invariant under $e^{i\theta}$ and it is a null space since $\mathbf{M} \cap \mathbf{M}'$ is null and $[e^{ik\theta}(\mathbf{M} \cap \mathbf{M}'), e^{iq\theta}(\mathbf{M} \cap \mathbf{M}')] = [e^{i(k-q)\theta}(\mathbf{M} \cap \mathbf{M}'), (\mathbf{M} \cap \mathbf{M}')] = 0$. One can check that $\mathbf{N}$ is maximal negative in $\mathbf{M}$. By the proof of Theorem 5.A.1 we see that the angle operator for $\mathbf{N}$ is a function f in $\overline{\mathbf{B}}H^{\infty;l}$ which solves the original interpolation problem. That $\mathbf{N}$ is null says that its angle operator is isometric which says that $|f(e^{i\theta})| = 1$ for a.e. θ. Consequently, we have shown that an inner solution of the interpolation problem exists.

It is reassuring that f is very easy to compute explicitly from our "geometric" picture. Here is how when $l = 0$. Solve $\Lambda_{\{z,w\}} x = 0$ for $x = (x_1, \ldots, x_p)^T$ and note from (5.3) that $\binom{u}{v} = \sum_{i=1}^{p} x_i b_i$. Assuming that the null space of Λ has dimension 1, which generally is the case, $\mathbf{M} \cap \mathbf{M}'$ is one-dimensional, so $\binom{u}{v}$ is a basis for it. Then $\mathbf{N} = \{ \binom{uh}{vh} : h \in H^\infty \}$ and the angle operator F for $\mathbf{N}$ satisfies $Fvh = uh$. Thus $f = u/v = [\sum_{i=1}^{p} x_j k_{z_i}][\sum_{i=1}^{p} x_i \bar{w}_i k_{z_i}]^{-1}$.

B. Toeplitz Corona problem. *Given* $a_1, a_2 \in H^\infty$, *given* $\kappa > 0$, *and given* l *an integer, find*

$$\mathbf{C}_\kappa^l \triangleq \left\{ \binom{h_1}{h_2} \in \overline{\mathbf{B}}H_{2\times1}^\infty : \kappa\phi \triangleq a_1 h_1 + a_2 h_2 \text{ is rational,} \right.$$

$$\left. \text{has } |\phi(e^{i\theta})| \equiv 1, \text{ and has winding number about zero } \leq l \right\}.$$

Abbreviate winding number about zero to wno. The classical Corona problem has $l = 0$. Then $\kappa\phi \triangleq a_1 h_1 + a_2 h_2$ is in H^∞ since a_j and h_j are; so wno $\phi \geq 0$. However, wno $\phi \leq l = 0$ implies wno $\phi = 0$. Thus ϕ is a constant which we can divide by κ to obtain that $\mathbf{C}_\kappa^0$ is nonempty if and only if there is an $\binom{\tilde{h}_1}{\tilde{h}_2}$ in $(1/\kappa)\overline{\mathbf{B}}H_{2\times1}^2$ solving

(BI) $$a_1 \tilde{h}_1 + a_2 \tilde{h}_2 = 1.$$

Here $\tilde{h}_j = h_j/\kappa$. What we have just cited is the familiar form of the Corona problem. If the reader finds the extra freedom allowed by $l \neq 0$ confusing, he can ignore it through most of this exposition. However, we assure the reader

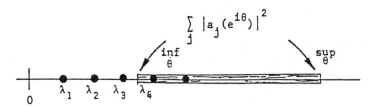

FIGURE 5.1. Spectrum of τ.

that this extra generality is canonical and that we ultimately need something even more general to study the optimization problem (OPT) of Chapter 2.

Historically each a_j is continuous and we want continuous solutions $\tilde{h}_j$ to (BI) without regard to the size of $\tilde{h}_j$. This was solved by Wiener whose famous Tauberian theorem says that a solution exists if no z_0 is a zero of all the a_j simultaneously. Gelfand theory trivialized the proof of Wiener's theorem. A far harder problem is merely to assume that a_j is in H^∞ and try to find if some $\tilde{h}_j$ in H^∞ satisfies (BI). This was solved by

CARLESON'S CORONA THEOREM. *If*

$$\inf_{|z|<1} |a_1(z)|^2 + |a_2(z)|^2 > \delta > 0,$$

then there exist $\tilde{h}_j$ *in* H^∞ *which solve* (BI).

What makes the problem hard is that from knowledge of δ one must compute an a priori upper bound on the size of the $\tilde{h}_j$. Even though it need not be a precise upper bound, computing any upper bound even when the a_j are continuous is difficult. Consequently it is surprising that the following theorem is relatively easy to prove.

THEOREM 5.B.1. *Set*

$$\tau = T_{a_1}T_{a_1}^* + T_{a_2}T_{a_2}^*;$$

then the Toeplitz Corona problem has a solution if the $l + 1$*st eigenvalue of* τ *is* $> \kappa^2$ *and only if it is* $\geq \kappa^2$. *Here continuous spectrum counts as infinitely many eigenvalues (see Figure* 5.1).

This theorem for $l = 0$ was discovered by Nagy and Foiaş [NF2] and independently, but later, by several people including M. Rosenblum [R] and W. Arveson [A]. It is an immediate consequence of the commutant lifting theorem [NF1]. A strong relationship between the Toeplitz Corona Theorem and the classical one follows from

$$([a_1 P_{H^2}\bar{a}_1 + a_2 P_{H^2}\bar{a}_2]f, f)_{L^2} \geq \kappa^2 (P_{H^2}f, f)_{L^2}.$$

It says the integral operator with kernel

$$\frac{a_1(z)\overline{a_1(w)} + a_2(z)\overline{a_2(w)}}{1 - z\bar{w}}$$

is "bigger" than the integral operator with kernel $\kappa^2 1/(1 - z\bar{w})$. This is because P_{H^2} is the integral operator with kernel $1/(1-z\bar{w})$. Carleson's hypothesis implies that the diagonal entries are $\geq \delta(1 - |z|^2)^{-1}$, so clearly $\delta \geq \kappa^2$. Thus the classical Corona theorem amounts to getting a (crude) lower bound on κ^2 with only knowledge of δ, $\|a_1\|_{L^\infty}$, and $\|a_2\|_{L^\infty}$.

In this subsection we show how to approach the Toeplitz Corona problem using invariant subspaces and projective techniques. As we shall see, the proof follows the same pattern as the previous one for interpolation.

Suppose $a_1, a_2 \in H^\infty$ and to avoid technical problems assume that they are rational. As before, the key is to find subspace $\mathbf{M}$ to associate with a_1 and a_2. We set everything in the space H_3^2, which is a Krein space with respect to the sesquilinear form $[\ ,\ ]$ induced by the form $[\ ,\ ]_{\mathbb{C}^3}$ on $\mathbb{C}^3$ given by

$$[x, y]_{\mathbb{C}^3} = x_1 \bar{y}_1 + x_2 \bar{y}_2 - x_3 \bar{y}_3,$$

and with respect to the decomposition $\mathbf{K}_+ = \left(\begin{smallmatrix} H^2 \\ 0 \end{smallmatrix}\right)$ and $\mathbf{K}_- = \left(\begin{smallmatrix} 0 \\ H^2 \end{smallmatrix}\right)$. Take $\mathbf{M}$ to be

$$\mathbf{M} = \left\{ \begin{pmatrix} u_1 \\ u_2 \\ \frac{a_1}{\kappa} u_1 + \frac{a_2}{\kappa} u_2 \end{pmatrix} : \begin{pmatrix} u_1 \\ u_2 \end{pmatrix} \in H_2^2 \right\}.$$

If $\mathbf{M} \cap \mathbf{M}' = \{0\}$ and $\mathbf{M} + \mathbf{M}' = L_3^2$, there exists g, a 3×2 matrix-valued function, so that $\mathbf{M} = g H_2^2$. Moreover, $g(e^{i\theta})$ is an isometry from a sesquilinear form $[\ ,\ ]_{\mathbb{C}^2}$ on $\mathbb{C}^2$ to $[\ ,\ ]_{\mathbb{C}^3}$ on $\mathbb{C}^3$. Indeed we can take $[\ ,\ ]_{\mathbb{C}^2}$ to be the standard form of either signature $1, 1$ or $2, 0$ and $\mathbb{C}^2$ depending on the signature of $[\ ,\ ]$ restricted to $\mathbf{L} \triangleq e^{i\theta} \mathbf{M}' \cap \mathbf{M}$. We shall soon see that the hypothesis of Theorem 5.B.1 forces the signature to be $1, 1$. Thus $g = \left(\begin{smallmatrix} \alpha & \beta \\ \kappa & \gamma \end{smallmatrix}\right)$ produces a L.F.T., $\mathbf{G}_g(s) = (\alpha s + \beta)(\kappa s + \gamma)^{-1}$, which takes 1×1 matrix functions s to 2×1 matrix functions. Here α, β are 2×1 matrices while κ, γ are 1×1 matrices.

THEOREM 5.B.2. *Suppose $\mathbf{M}' + \mathbf{M} = L_3^2$ and $\mathbf{M} \cap \mathbf{M}' = \{0\}$ (this happens if and only if $\tau - \kappa^2$ is invertible). Set $l =$ the negative signature of $\mathbf{M}' \cap H_3^2$. Then $\mathbf{C}_\kappa^l = \mathbf{G}_g(\overline{\mathbf{B}} H^\infty)$. When $l' < l$ the set $\mathbf{C}_\kappa^{l'}$ is empty. Moreover, the integer l can be computed from τ as in Theorem 5.B.1.*

PROOF. This is an exact repetition of the interpolation proof; enough so that it should not be repeated. However, for pedantic effect, we go through the details.

(i) $\mathbf{C}_\kappa^l$ is the set of all angle operators for invariant negative subspaces of $\mathbf{M}$ which have negative cosignature l in H_3^2.

By Fact 3.C.4 a negative invariant subspace $\mathbf{S}$ has an angle operator $H : \left(\begin{smallmatrix} 0 \\ D \end{smallmatrix}\right) \subset \left(\begin{smallmatrix} 0 \\ H^2 \end{smallmatrix}\right) \to \left(\begin{smallmatrix} H^2 \\ 0 \end{smallmatrix}\right)$ which is multiplication by a function $\left(\begin{smallmatrix} h_1 \\ h_2 \end{smallmatrix}\right)$ in $\overline{\mathbf{B}} H_2^{\infty, l}$.

Now $\mathbf{S} \subset \mathbf{M}$ says precisely that for any u in D,

$$\text{the function } \begin{pmatrix} h_1 u \\ h_2 u \\ u \end{pmatrix} \text{ has the form } \begin{pmatrix} u_1 \\ u_2 \\ \frac{a_1}{\kappa} u_1 + \frac{a_2}{\kappa} u_2 \end{pmatrix}$$

for some pair $\begin{pmatrix} u_1 \\ u_2 \end{pmatrix}$ in H_2^2. Thus $h_1 u = u_1$, and $h_2 u = u_2$, and $a_1 h_1 u + a_2 h_2 u = \kappa u$. Since, by the classical Beurling-Lax Theorem, D can be represented as $D = \phi H^2$ for some inner ϕ, the functions $h_1 \phi$ and $h_2 \phi$ are both in H^∞. Consequently

$$a_1 \tilde{h}_1 + a_2 \tilde{h}_2 = \kappa \phi,$$

where $\tilde{h}_j = h_j \phi \in H^\infty$. Moreover, $\|\tilde{H}\|_{L_{2 \times 1}^\infty} = \|H\|_{L_{2 \times 1}^\infty} \leq 1$. So $\begin{pmatrix} \tilde{h}_1 \\ \tilde{h}_2 \end{pmatrix} \in \mathbf{C}_\kappa^l$. That is, each $\mathbf{S}$ produces an element of $\mathbf{C}_\kappa^l$. Work backward through the argument to get the converse.

(ii)   *The invariant maximal negative subspaces of $\mathbf{M} = g($the invariant maximal negative subspaces of $H_2^2)$.*

(iii)   *Each invariant maximal negative subspace of $\mathbf{M}$ has negative cosignature l in H_2^2 where $l =$ the negative signature of $\mathbf{M}' \cap H_2^2$.*

Combine (i), (ii), and (iii).

$\mathbf{C}_\kappa^l$ is the set of all angle operators for g (the invariant maximal negative subspaces of H_2^2).

This is the geometric statement of the main part of Theorem 5.B.2. At this stage the second part of Theorem 5.B.2 amounts to

THEOREM 5.B.3. *The negative signature of $\mathbf{M}' \cap H_2^2$ equals the number of negative eigenvalues of $\tau - \kappa^2$.*

PROOF. A vector v in H_3^2 is in $\mathbf{M}'$ if and only if

$$0 = \left[\begin{pmatrix} v_1 \\ v_2 \\ v_3 \end{pmatrix}, \begin{pmatrix} u_1 \\ u_2 \\ \frac{a_1}{\kappa} u_1 + \frac{a_2}{\kappa} u_2 \end{pmatrix} \right] = (v_1, u_1) + (v_2, u_2) - \left(\frac{\overline{a}_1}{\kappa} v_3, u_1 \right) - \left(\frac{\overline{a}_2}{\kappa} v_3, u_2 \right)$$

for all $u_1, u_2 \in H^2$. This is equivalent to $v_j = T_{\overline{a}_j / \kappa} v_3$, so we have characterized $\mathbf{M}' \cap H_3^2$ as all vectors of the form

$$\begin{pmatrix} T_{\overline{a}_1} \rho \\ T_{\overline{a}_2} \rho \\ \kappa \rho \end{pmatrix}$$

for ρ in H^2. Consequently, $[\ ,\ ]$ evaluated on two vectors in $\mathbf{M}' \cap H_3^2$ is

$$\left[\begin{pmatrix} T_{\overline{a}_1} \rho \\ T_{\overline{a}_2} \rho \\ \kappa \rho \end{pmatrix}, \begin{pmatrix} T_{\overline{a}_1} \dot{\rho} \\ T_{\overline{a}_2} \dot{\rho} \\ \kappa \dot{\rho} \end{pmatrix} \right] = (\{T_{\overline{a}_1}^* T_{\overline{a}_1} + T_{\overline{a}_2}^* T_{\overline{a}_2} - \kappa^2\} \rho, \dot{\rho})_{H^2}$$

$$= (\{\tau - \kappa^2\} \rho, \dot{\rho})_{H^2}.$$

The statement about negative signature follows immediately. Moreover, $\mathbf{M}' \cap \mathbf{M} \neq \{0\}$ if and only if $\tau - \kappa^2$ has a null vector. One can also show that $\mathbf{M} + \mathbf{M}' = L^2$ if and only if $\tau - \kappa^2$ is invertible.

C. Factorization. Wiener-Hopf factorization was already treated in Chapter 4 where we saw that it is equivalent to the invariant subspace theorems we have been using repeatedly. The point of bringing it up again (besides as a reminder) is to indicate some refinements and variants. In these we try to obtain factorizations where the factors meet certain restrictions. Here we concentrate on restrictions connected with a signed sesquilinear form.

Again define $[\,,\,]_{\mathbf{C}^N} = x_1 \bar{y}_1 + \cdots + x_m \bar{y}_m - x_{m+1} \bar{y}_{m+1} - x_{m+n} \bar{y}_{m+n}$. Associated with it is $J = \begin{pmatrix} I_m & 0 \\ 0 & -I_n \end{pmatrix}$. Let A^+ denote the adjoint for the matrix $A : \mathbf{C}^N \to \mathbf{C}^K$ with respect to $[\,,\,]_{\mathbf{C}^N}$ and $[\,,\,]_{\mathbf{C}^K}$, that is, $[Ax, y]_{\mathbf{C}^K} = [x, A^+ y]_{\mathbf{C}^N}$ all x, y. Naturally $A^+ = A$ says that A is "*J-selfadjoint*" while an A satisfying $A^+ J_K A = J_N$ is "*J-isometric.*"

Wiener-Hopf factorization when applied to a selfadjoint matrix function A (conventional adjoint here) yields

THEOREM 5.C.1. *If A and A^{-1} in $L^\infty_{N \times N}$ are continuous, and have selfadjoint values, then there exists Q in $H^2_{N \times N}$ such that*

(a) $A = Q^* J Q$ *where $J = \begin{pmatrix} I_m & 0 \\ 0 & -I_n \end{pmatrix}$ has the same signature as does $A(e^{i\theta})$ for a.e. θ. For generic A we may take Q outer.*

(b) $A = Q^* D Q$ *with Q outer and D a (unique) Hermitian winding matrix (as defined below). The winding matrix is independent of θ if and only if T_A is invertible.*

By a *Hermitian winding matrix* $D(e^{i\theta})$ we mean a variant on the winding matrix of Chapter 4.

$$D(e^{i\theta}) = \begin{bmatrix} & & & & & & & & e^{ik_1\theta} \\ & & & & & & & e^{ik_2\theta} & \\ & & & & & & \ddots & & \\ & & & & & e^{ik_r\theta} & & & \\ & & & & \begin{bmatrix} I_{\alpha_+} & 0 \\ 0 & -I_{\alpha_-} \end{bmatrix} & & & & \\ & & & e^{-ik_r\theta} & & & & & \\ & & \ddots & & & & & & \\ & e^{-ik_2\theta} & & & & & & & \\ e^{-ik_1\theta} & & & & & & & & \end{bmatrix}.$$

This theorem is due to Nikoliachuk and Spitskovski [**NS1-NS2**], see also Chapter V of Clancy and Gohberg [**CG**] or [**BH8**] for an invariant subspace proof of the theorem. When A is positive definite one has the traditional

COROLLARY 5.C.2 (WIENER-MASANI). *If A is also positive definite, then $\exists Q$ outer such that $A = Q^* Q$.*

PROOF OF THEOREM 5.C.1. Let $\mathbf{K}$ be the Krein space consisting of the functions in the weighted L^2 space $L_N^2(|A|)$ with the sesquilinear form

$$[f, h] = \frac{1}{2\pi} \int (Af(e^{i\theta}), h(e^{i\theta}))_{\mathbb{C}^N} \, d\theta.$$

Since A and A^{-1} are uniformly bounded $\mathbf{K}$, $[\,,\,]$ is indeed a Krein space. Define $\mathbf{M}$ in $L_N^2(|A|)$ to be

$$\mathbf{M} = \text{the closed span of } (e^{in\theta} \text{ constants}).$$

It is simply invariant under $M_{e^{i\theta}}$.

The representation Theorem 4.1 applied to $\mathbf{M}$ and $\mathbf{M}'$ implies that $\mathbf{M} = \text{clos } g(DH_N^\infty)$ and $\mathbf{M}' = \text{clos } \overline{gH}_N^\infty$ for a $g^{\pm 1} \in L_{N \times N}^2$ with $g(e^{i\theta}, \,)$ a $[\,,\,]_{\mathbb{C}^N} \to (A(e^{i\theta}, \,)_{\mathbb{C}^N}$ isometry. The prerequisite $\mathbf{M} \cap \mathbf{M}'$ finite dimensional and $\mathbf{M} + \mathbf{M}'$ cofinite dimensional follows from continuity of A (since it implies T_A is Fredholm). Moreover, $D = I$ if and only if T_A is invertible.

When $D = I$, we have that $g^*Ag = \begin{bmatrix} I & 0 \\ 0 & -I \end{bmatrix}$, and so $A = g^{*-1}\begin{bmatrix} I & 0 \\ 0 & -I \end{bmatrix}g^{-1}$. Since $\text{clos } gH_N^2 = \mathbf{M} = H_N^2(|A|)$ we get that g and g^{-1} are in $H_{N \times N}^2$. Thus $Q = g$ satisfies (b). Also Theorem 4.1 implies $g^{-1}P_{H_N^2}g$ is bounded on L^2. When $D \neq I$, then generalizing this argument is straightforward and one obtains (b).

Part (a) is true because $D(e^{i\theta})$ factors as $Q_1^*JQ_1$, for example when $N = 2$ and $m = n = 1$ where

$$Q_1 = \frac{1}{\sqrt{2}} \begin{pmatrix} 1 & 1 \\ 1 & -1 \end{pmatrix} \begin{pmatrix} 1 & 0 \\ 0 & e^{ik_1\theta} \end{pmatrix}.$$

A variant on all this is inner-outer factorization.

THEOREM 5.C.3. *Suppose A is in $H_{K \times N}^\infty$, is continuous, and its values are full rank matrices. Here $N \leq K$. If J is a signature matrix on $\mathbb{C}^K$, then there is an outer Q in $H_{N \times N}^2$ and a U in $H_{N \times N}^2$ such that $A = UQ$ and $U^*JU = D$ for some Hermitian winding matrix D. Naturally D is determined uniquely. If $J = 1$, then U is inner. For generic A the winding matrix D is constant.*

THEOREM 5.C.4 (LEECH [L]). *If $A \in L_{N \times K}^\infty$ and $F \in L_{N \times R}^\infty$, then there is a $G \in H_{K \times R}^\infty$ for which*

$$F = AG$$

if and only if $FH_R^2 \subset AH_K^2$. Moreover, $\|G\|_{L^\infty} \leq 1$ if and only if $FF^* \leq AA^*$.

Proof of Theorems 5.C.3 and 5.C.4 are left as exercises.

D. Disk problems. *Parametrize $\mathbf{A} \cap \Delta_K^R$. Is it empty?*

Since engineering calls for it we shall operate at a high level of generality. A disk in the matrix function space $L_{m \times n}^\infty$ is a set of the form

$$\Delta_K^{P,R} = \{F \in L_{m \times n}^\infty : [F(e^{i\theta}) - K(e^{i\theta})]P(e^{i\theta})^2[F(e^{i\theta}) - K(e^{i\theta})]^* \leq R(e^{i\theta})^2\}$$

where P^2 and R^2 are positive selfadjoint matrix-valued functions.

THEOREM 5.D.1. *If $P^2 \geq \varepsilon_1 I$ and $R^2 \geq \varepsilon_2 I$ are rational, then one can construct a finite-dimensional matrix Γ with the properties*

(1) $\Gamma \not\geq 0 \Leftrightarrow \mathbf{A}_{m \times n} \cap \Delta_K^{P,R}$ *is empty.*

(2) $\Gamma > 0 \Leftrightarrow \mathbf{A}_{m \times n} \cap \Delta_K^{P,R}$ *is not empty and has a L.F.T. parametrization* $\mathbf{G}_g(\overline{\mathbf{B}}H_{m \times n}^\infty)$.

(3) Γ *has l negative eigenvalues $\Leftrightarrow H_{m \times n}^{\infty;l} \cap \Delta_K^{P,R}$ is not empty.*

This theorem is less general than what one could prove and indeed is a special case of A in Chapter 6. We call special attention to it by creating this separate section simply because this theorem is the one which is used to solve the engineering problem in Chapter 2. Of the proofs one could give, we outline two (see also Chapter 6.A).

PROOF. Let α and β be the outer Wiener-Hopf factors

$$\beta\beta^* = P^2 \quad \text{and} \quad \alpha\alpha^* = R^2.$$

Since $P^2 > \varepsilon_1 I$ and $R^2 > \varepsilon_2 I$ we have that $\alpha, \alpha^{-1}, \beta, \beta^{-1}$ are all in matrix-valued H^∞. Set $H = \alpha^{-1}F\beta$. Then $F \in \Delta_K^{P,R}$ if and only if

$$(H - \alpha^{-1}K\beta)(H - \alpha^{-1}K\beta)^* \leq I.$$

As F sweeps through $\mathbf{A}_{m \times n}$, the function H sweeps through $\mathbf{A}_{m \times n}$ also. Thus

$$\mathbf{A}_{m \times n} \cap \Delta_K^{P,R} \text{ corresponds to } \mathbf{A}_{m \times n} \cap \Delta_{\alpha^{-1}K\beta}^{I,I}$$

via the map $F \to \alpha^{-1}F\beta$. If we set $K_1 = \alpha^{-1}K\beta$, then our main problem becomes one of parametrizing

$$\mathbf{W} \triangleq \{H \in \mathbf{A}_{m \times n} \colon \|H - K_1\|_{L_{m \times n}^\infty} \leq 1\}.$$

Now we give two ways of doing this.

The first is to convert this to an interpolation problem. Let ϕ be in $\mathbf{A}_{n \times n}$, have unitary values on the unit circle, and let it force $K_1\phi$ to be in $\mathbf{A}_{m \times n}$. That is, the "zeros" of ϕ cancel the "poles" of K_1 inside the unit disk. Let

$$\mathbf{T} = \{Y \in \mathbf{A}_{m \times n} \colon Y = H\phi - K_1\phi \text{ for some } H \text{ in } \mathbf{W}\}.$$

The set $\mathbf{T}$ is an interpolation type of set like those studied in A of Chapter 5. The only difference is that we have a matrix (rather than scalar) version of such a set. To see this think only of the scalar case and let z_j be the zeros of ϕ, which, of course, are the poles of K_1. Set $w_j = K_1\phi(z_j)$. Then

$$Y \in \mathbf{T} \Leftrightarrow Y \in \mathbf{A} \text{ and } Y(z_j) = w_j,$$

and so we have reduced our "disk" problem to one solved by the method in A of Chapter 5. This yields Theorem 5.D.1.

The second proof is simply the direct one. To the set $\mathbf{W}$ we associate a subspace $\mathbf{M}$ of $L_{m \times n}^2$ and then proceed as before. The selection of $\mathbf{M}$ and details are left as a homework exercise. If it proves troublesome see Chapter 6.A.

E. Appendix: More on interpolation. This section consists of miscellany on interpolation, the Corona theorem, and factorization.

Although it is redundant we give a small table which summarizes the proof of interpolation Theorem 5.A.1.

Table 5.E.1. Proof of Theorem 5.A.1 for $l = 0$.

$$\mathbf{K} = H_2^2 \text{ with } [x, y]_{\mathbb{C}^2} = x_1 \bar{y}_1 - x_2 \bar{y}_2, \quad \mathbf{M} = \{\begin{pmatrix} u \\ v \end{pmatrix} \in H_2^2 : u(z_j) = w_j v(z_j)\}.$$

(i) $\left. \begin{array}{ll} \mathbf{S} \text{ invar. max. neg. in } H_2^2 & \Leftrightarrow \quad \text{angle op. } f \in \overline{\mathbf{B}}H^\infty \\ \mathbf{S} \text{ also in } \mathbf{M} & \Leftrightarrow \quad f(z_j) = w_j \end{array} \right\} \Leftrightarrow f \in \mathbf{T}^0.$

(ii)   Every $\mathbf{S}$ invar. max. neg. in $\mathbf{M} = g(\text{invar. max. neg. in } H_2^2)$.

(iii) $\left. \begin{array}{l} \mathbf{S} \text{ invar. max. neg. in } \mathbf{M} \\ \text{is also max. neg. in } H_2^2 \end{array} \right\} \Leftrightarrow \mathbf{M}' \cap H_2^2 \text{ is positive}$

Put it together to get

$$\mathbf{M}' \cap H_2^2 \text{ positive} \Rightarrow \{\mathbf{S}_f : f \in \mathbf{T}^0\} = g\{\mathbf{S}_f : f \in \overline{\mathbf{B}}H^\infty\}.$$

The summary table for the proof of Corona Theorem 5.B.2 is as follows:

Table 5.E.2. Proof of Theorem 5.B.2 for $l = 0$.

$$\mathbf{K} = H_3^2 \text{ with } [x, y]_{\mathbb{C}^3} = x_1 \bar{y}_1 + x_2 \bar{y}_2 - x_3 \bar{y}_3,$$

$$\mathbf{M} = \left\{ \begin{pmatrix} u \\ v \\ \frac{a_1 u}{\kappa} + \frac{a_2 v}{\kappa} \end{pmatrix} \in H_3^2 \right\}.$$

(i) $\left. \begin{array}{ll} \mathbf{S} \text{ invar. max. neg. in } H_2^2 & \Leftrightarrow \quad \text{angle op. } f \in \overline{\mathbf{B}}H_2^2 \\ \mathbf{S} \text{ also in } \mathbf{M} & \Leftrightarrow \quad a_1 f_1 / \kappa + a_2 f_2 / \kappa = 1 \end{array} \right\} \Leftrightarrow f \in \mathbf{C}_\kappa^0.$

(ii) $\left. \begin{array}{l} \text{Every } \mathbf{S} \text{ invar. max. neg. in } \mathbf{M} \\ \text{is also max. neg. in } H_3^2 \end{array} \right\} \Leftrightarrow \mathbf{M}' \cap H_3^2 \text{ positive.}$

(iii) $\mathbf{S}$ invar. max. neg. in $\mathbf{M} = g(\text{invar. max. neg. in } H_3^2)$.

Put it together to get

$$\mathbf{M}' \cap H_3^2 \text{ positive} \Rightarrow \{\mathbf{S}_f : f \in \mathbf{C}_\kappa^0\} = g\{\mathbf{S}_f : f \in \overline{\mathbf{B}}H^2\}.$$

Solution to Exercise 5.1. To compute g one merely must compute $\mathbf{L}$ in the proof of Theorem 4.1. So we now do this under the assumption that no interpolating condition is imposed at $z_j = 0$. By definition $\mathbf{L} = \mathbf{M} \boxminus S\mathbf{M} = \mathbf{M} \cap (S\mathbf{M})'$. Since

$$S\mathbf{M} = \left\{ \begin{pmatrix} u \\ v \end{pmatrix} \in H_2^2 : u(z_j) = w_j v(z_j), \ u(0) = v(0) = 0 \right\}$$

we see that $(S\mathbf{M})'$ equals $\{\binom{a}{b} \in \mathbb{C}^2\} + S\mathbf{M}'$, so it consists of all f of the form

$$f = \binom{a}{b} + \sum_{j=1}^{p} \alpha_j e^{i\theta} b_j$$

with $a, b, \alpha_j \in \mathbb{C}$. The functions in $\mathbf{L}$ satisfy the additional requirement

$$0 = [f, b_\nu] = \left[\binom{a}{b}, b_\nu\right] + \sum_{j=1}^{p} \alpha_j [e^{i\theta} b_j, b_\nu]$$

$$= a - w_\nu b + \sum_{j=1}^{p} \alpha_j z_\nu \frac{1 - \bar{w}_j w_\nu}{1 - \bar{z}_j z_\nu}$$

for $\nu = 1, 2, \ldots, p$. In terms of the Pick matrix Γ this says

$$0 = a\vec{1} - \vec{w}b + \begin{pmatrix} z_1 & 0 & & 0 \\ 0 & z_2 & & 0 \\ & & \ddots & \\ 0 & 0 & & z_p \end{pmatrix} \Gamma\vec{\alpha}$$

where

$$\vec{w} = \begin{pmatrix} w_1 \\ \vdots \\ w_p \end{pmatrix} \text{ and } \vec{\alpha} = \begin{pmatrix} \alpha_1 \\ \vdots \\ \alpha_p \end{pmatrix} \text{ and } \vec{1} = \begin{pmatrix} 1 \\ 1 \\ 1 \\ 1 \end{pmatrix}.$$

Thus a basis for $\mathbf{L}$ comes from solving

$$\Gamma\vec{\alpha}^1 = \begin{pmatrix} z_1^{-1} \\ \vdots \\ z_p^{-1} \end{pmatrix} \quad \text{and} \quad \Gamma\vec{\alpha}^2 = \begin{pmatrix} z_1^{-1}w_1 \\ \vdots \\ z_p^{-1}w_p \end{pmatrix}$$

for $\vec{\alpha}^1$ and $\vec{\alpha}^2$. Then set $\sigma^s(e^{i\theta}) = \sum_{j=1}^{p} \alpha_j^s b_j$ and obtain that

$$f^1(e^{i\theta}) = \binom{1}{0} - \sigma^1(e^{i\theta}), \qquad f^2(e^{i\theta}) = \binom{1}{0} + \sigma^2(e^{i\theta})$$

is a basis for $\mathbf{L}$.

To obtain the representer g for $\mathbf{M}$ and $\mathbf{M}'$ all that we need to do is $[\ ,\ ]$ Gram-Schmidt the basis f^1, f^2 to obtain λ^+ and λ^- $[\ ,\ ]$ orthogonal positive and negative vectors. Then

$$g = \begin{pmatrix} \lambda_1^+ & \lambda_1^- \\ \lambda_2^+ & \lambda_2^- \end{pmatrix}$$

where $\lambda^\pm = \binom{\lambda_1^\pm}{\lambda_2^\pm}$ is the representer. Why? Since $\lambda^\pm$ are a basis for $\mathbf{L}$ we see g(the constants)= $\mathbf{L}$. The orthogonality of λ^+ and λ^- means that g is a $[\ ,\ ]$ isometry.

6. Further Applications

This chapter presents more applications of our invariant subspace representation. While the applications presented in Chapter 5 are used in other parts of these notes, the applications presented in Chapter 6 are not critical to the rest of the chapters.

The items for presentation are:

A. A theorem which contains both interpolation and Toeplitz Corona theory.

B. Added symmetries.

C. Interpolation on the boundary.

D. Commutant lifting theory.

E. F. and M. Riesz Theorem.

F. Completely integrable differential equations.

Sections 6.A, 6.B, 6.C are set in exactly the same context as was Chapter 5; again we have L_N^2, H_N^2, **M**, **S**, etc.

A. Analytic systems. *Given $\psi, Y \in L^\infty$ and $a_1, a_2 \in H^\infty$, given an integer l, and given $\kappa > 0$, find $\mathbf{L}^l$ defined to be the set of all $\alpha \in \overline{\mathbf{B}}L^\infty$ and ϕ inner of degree $\leq l$ for which there exists G_1 and G_2 in H^∞ satisfying*

(AS)
$$\alpha = \psi G_1 + Y G_2,$$
$$\phi\kappa = a_1 G_1 + a_2 G_2.$$

All theorems in Chapter 5 are special cases of our forthcoming theorem about these systems.

(a) *Interpolation and disk problems.* Given $Y \in L^\infty$, $l \geq 0$, and ψ inner in H^∞, find all functions in

$$\mathbf{T}^l = \{\alpha \in Y + \psi H^{\infty;l} : \|\alpha\| \leq 1\}.$$

EXAMPLE 1. Set $\psi = 1$. Then $\mathbf{T}^l$ parametrizes all functions in $H^{\infty;l}$ whose distance to the given function Y is ≤ 1. This is the (generalized) Nehari problem, the most basic disk problem we studied in Chapter 5.D.

EXAMPLE 2. Given z_j, w_j; let $\psi \in H^\infty$ have zeros precisely at the z_j and let Y be any H^∞ function satisfying $Y(z_j) = w_j$. Then $\mathbf{T}^l$ equals $\{\alpha \in H^{\infty;l} : \alpha(z_j) = Y(z_j = w_j)\}$. Thus $\mathbf{T}^l$ equals $\mathbf{T}^l$ in the interpolation section in Chapter 5.A.

57

Conversion of $\mathbf{T}^l$ *to* $\mathbf{L}^l$. For simplicity of explanation take $l = 0$. Take $a_1 = 0$, $a_2 = 1$, and $\kappa = 1$. If $\left(\begin{smallmatrix} \alpha \\ \phi \end{smallmatrix}\right) \in \mathbf{L}^0$, then $\phi = 1$ and $\exists G_1, G_2 \in H^\infty$ so that

(6.1) $\alpha = \psi G_1 + Y G_2, \qquad 1 = G_2.$

Thus

$$Y + \psi G_1 = \alpha,$$

which has $\| \ \|_{L^\infty} \leq 1$, that is, $\alpha \in \mathbf{T}^0$. Conversely an α in $\mathbf{T}^0$ produces an $\left(\begin{smallmatrix} \alpha \\ 1 \end{smallmatrix}\right)$ in $\mathbf{L}^0$.

(b) *Corona.* See Chapter 5.B. Take $\psi = \left(\begin{smallmatrix} 1 \\ 0 \end{smallmatrix}\right)$, $Y = \left(\begin{smallmatrix} 0 \\ 1 \end{smallmatrix}\right)$. Then (AS) becomes

$$\alpha = \begin{pmatrix} G_1 \\ G_2 \end{pmatrix}, \qquad \kappa\phi = a_1 G_1 + a_2 G_2.$$

So $\left(\begin{smallmatrix} \alpha \\ \phi \end{smallmatrix}\right) \in \mathbf{L}^l$ if and only if $\left(\begin{smallmatrix} G_1 \\ G_2 \end{smallmatrix}\right) \in \mathbf{C}_K^l$.

Before giving results on analytic systems we remark that all functions involved can be taken to be matrix-valued. To wit, take

$$\psi \in L_{m \times t}^\infty, \qquad Y \in L_{m \times s}^\infty, \qquad a_1 \in H_{n \times t}^\infty, \qquad a_2 \in H_{n \times s}^\infty,$$
$$\alpha \in \overline{\mathbf{B}} L_{m \times n}^\infty, \qquad \phi \text{ inner in } H_{n \times n}^\infty, \qquad G_1 \in H_{t \times n}^\infty, \qquad G_2 \in H_{s \times n}^\infty.$$

Then problem (AS) still makes perfect sense, and the specialization of it to interpolation and Corona problems simply gives the matrix version of them. Also all of our results still hold, so the proofs which follow are given for matrix-valued functions.

In studying problem (AS) the matrix function

$$A = \begin{pmatrix} \psi & Y \\ \dfrac{a_1}{\kappa} & \dfrac{a_2}{\kappa} \end{pmatrix} \quad \text{in } L_{N \times K}^\infty$$

plays a major role. Here $N = m + r$ and $K = t + s$. The signature matrix $J = \left(\begin{smallmatrix} I_m & 0 \\ 0 & -I_n \end{smallmatrix}\right)$ gives a sesquilinear form $[\ ,\ ]_{\mathbf{K}}$ on $\left(\begin{smallmatrix} L_m^2 \\ H_n^2 \end{smallmatrix}\right) \triangleq \mathbf{K}$; here

$$\mathbf{K}_+ = \begin{pmatrix} L_m^2 \\ 0 \end{pmatrix} \quad \text{and} \quad \mathbf{K}_- = \begin{pmatrix} 0 \\ H_n^2 \end{pmatrix}.$$

As usual there is a key invariant subspace $\mathbf{M} = A H_K^2 \subset \mathbf{K}$. If $\mathbf{M} + (\mathbf{M}' \cap \mathbf{K}) = \mathbf{K}$ and $\mathbf{M} \cap \mathbf{M}' = \{0\}$ we represent

$$\mathbf{M} = g H_K^2,$$

where $g(e^{i\theta})$ is a $[\ ,\ ]_{\mathbf{C}^K}$ to $(J,\ )_{\mathbf{C}^N}$ isometry a.e. Here $[\ ,\ ]_{\mathbf{C}^K}$ is a fixed indefinite sesquilinear form on $\mathbb{C}^K$. Familiar constructions yield for $K \leq N$

THEOREM 6.A.1. *Suppose* $T_{A^* J A}$ *is invertible. Let* l *be the negative signature of the sesquilinear form* $[\ ,\ ]$ *restricted to*

$$\left\{ \eta \in \begin{pmatrix} L_m^2 \\ H_n^2 \end{pmatrix} : P_{H_K^2} A^* J \eta = 0 \right\}.$$

Then all solutions $\binom{\alpha}{\delta}$ to problem (AS) *are of the form* $\alpha\delta^{-1} = \mathbf{G}_g(\overline{\mathbf{B}}H^\infty_{t\times s})$.

PROOF. One can use the familiar outline to prove this, but since it is so familiar we take a different tack.

Let $J = \left(\begin{smallmatrix} I_m & 0 \\ 0 & -I_m \end{smallmatrix}\right)$ and assume $A^*JA(e^{i\theta})$ has constant signature t, s almost everywhere in θ. Suppose (as in Theorem 5.C.1) that we can factor

$$A^*JA = Q^*\tilde{J}Q \quad \text{where } \tilde{J} = \begin{pmatrix} I_t & 0 \\ 0 & -I_s \end{pmatrix} \text{ and } Q \in H^2_{K\times K}$$

with $K = t + s$ and $Q^{-1} \in L^2_{K\times K}$. Set $g = AQ^{-1}$. Then $g^*Jg = \tilde{J}$ and so $g(e^{i\theta})$ is $[\,,\,]_{\mathbf{C}^K} \triangleq (\tilde{J}, \,)_{\mathbf{C}^K}$ to $(J, \,)_{\mathbf{C}^N}$ isometric. Set $\mathbf{R} = \text{clos } QH^2_K$.

The main point is that g *represents* $\mathbf{M}$ *in terms of the "reference" space* $\mathbf{R}$. Namely, $\mathbf{M} = \text{clos } g\mathbf{R}$. If Q is outer, then $\mathbf{R} = H^2_K$ and we are in the standard Beurling-Lax-Halmos situation which has pervaded the paper. The hypothesis T_{A^*JA} invertible via Theorem 5.C.1 is equivalent to the existence of Q outer, so that is why we indeed have $\mathbf{M} = gH^2_K$.

We reiterate that Q need not be outer; if one understands $\mathbf{R}$ well enough, then the representation $\mathbf{M} = \text{clos } g\mathbf{R}$ might be useful.

From arguments (i), (ii), and (iii) in the interpolation and Corona sections before, it is clear that if $\binom{\alpha}{\phi} \in \mathbf{L}^l$, then $\mathbf{S} = \binom{\alpha}{\phi}H^2_n$ is a maximal negative subspace of $\mathbf{M}$. Conversely, by Fact 3.C.4 such an $\mathbf{S}$ has an angle operator

$$F: \begin{pmatrix} 0 \\ D \end{pmatrix} \subset \begin{pmatrix} 0 \\ H^2_n \end{pmatrix} \to \begin{pmatrix} L^2_m \\ 0 \end{pmatrix}$$

which could be written as multiplication by $\binom{\alpha}{\phi}$ where $D = \phi H^2_n$ with ϕ inner. Since $\mathbf{S} \subset \mathbf{M}$, the range of $\binom{\alpha}{\phi}H^2_n$ is contained in the range of $\mathbf{A}$ acting on H^2_K. Consequently[1] $\exists G \in H^\infty_{K\times n}$ such that $\binom{\alpha}{\phi} = \mathbf{A}G$. We have proved that

$$\begin{pmatrix} \alpha \\ \phi \end{pmatrix} \in \mathbf{L}^l \Leftrightarrow \mathbf{S} \text{ is invariant maximal negative in } \mathbf{M}.$$

This is equivalent to the parametrization of $\mathbf{L}^l$ as $\mathbf{G}_g(\overline{\mathbf{B}}H^\infty_{t\times n})$.

The computation of l goes as follows. It is by definition given in terms of a maximal negative subsapce $\mathbf{S}$ of $\mathbf{M}$. Such a space is in $\mathbf{K}$ and so has a represent-tion $\binom{\alpha}{\phi}$ by Fact 3.C.4′. The function ϕ is inner in $H^\infty_{n\times n}$ and $\alpha \in \overline{\mathbf{B}}L^\infty_{m\times n}$. The integer l we seek is the degree of ϕ and by Fact 3.C.5, the negative signature of $\mathbf{M}' \cap \mathbf{K}$ equals l. To compute $\mathbf{M}' \cap \mathbf{K}$ consider η in $\mathbf{K}$ with

$$[\eta, \mathbf{A}H^2_{\mathbf{K}}]_{\mathbf{C}^N} = 0.$$

This says $P_{H^2_K}A\eta = 0$. The theorem is proved.[1]

B. Added symmetries. The key theorem in these lectures represents a pair $\mathbf{M}, \mathbf{M}^x$ of subspaces using a matrix-valued function g. The main applications in

---

[1]Even if $K > N$, a G will exist by Leech's theorem. Thus if $K \geq N$ one still gets Theorem 6.1 provided that A^*JA has an outer J factorization Q^*JQ.

Chapter 5 were based on the special case when $g(e^{i\theta})$ is in the Lie group $U(m,n)$. (Actually the Toeplitz Corona theorem is a variant on this.) What about the other classical matrix Lie groups?

Joe Ball and I in [**BH3**] and [**BH4**] characterized those **M** and **M**^{x} which can be represented by g in a particular classical Lie group. Then we gave applications in the spirit of Chapter 5. The motivation from engineering was a power transfer problems whose solution forced a study of $\text{Sp}(n,R)$. With $\text{Sp}(n,R)$ and $U(m,n)$ in hand we went on to do the other classical Lie groups. We now give a brief summary of results which follows the representation in [**BH4**] closely.

The classical matrix Lie groups are most easily defined in terms of involutions. Let $g \to g^x$ be an involutive automorphism on $\text{GL}(n,\mathbb{C})$; that is

(i) $(g_1 g_2)^x = g_1^x g_2^x$,

(ii) $(g_1^x)^x = g_1$,
         for all $g_1, g_2 \in \text{GL}(n,\mathbb{C})$. Let us say that the involutive auto-
         morphism ^{x} is Type I if

(iii.I) $F^x \in \overline{H_{n\times n}^{\infty}}$ whenever $F \in H_{n\times n}^{\infty}$ and $F^x \in H_{n\times n}^{\infty}$ whenever
          $F \in \overline{H_{n\times n}^{\infty}}$, and is Type II if

(iii.II) $F^x \in H_{n\times n}^{\infty}$ whenever $F \in H_{n\times n}^{\infty}$ and $F^x \in \overline{H_{n\times n}^{\infty}}$ whenever
          $F \in \overline{H_{n\times n}^{\infty}}$.

The main examples of Type I involutive automorphisms are

$$g \to \begin{bmatrix} I_p & 0 \\ 0 & -I_{n-p} \end{bmatrix} g^{*-1} \begin{bmatrix} I_p & 0 \\ 0 & -I_{n-p} \end{bmatrix}, \quad g \to \overline{g}$$

and

$$g \to \begin{bmatrix} 0 & I_m \\ -I_m & 0 \end{bmatrix} \overline{g} \begin{bmatrix} 0 & -I_m \\ I_m & 0 \end{bmatrix};$$

the main examples of Type II involutive automorphisms are

$$g \to g^{T-1}, \quad g \to \begin{bmatrix} 0 & I_m \\ -I_m & 0 \end{bmatrix} g^{T-1} \begin{bmatrix} 0 & -I_m \\ I_m & 0 \end{bmatrix},$$

and

$$g \to \begin{bmatrix} I_p & 0 \\ 0 & -I_{n-p} \end{bmatrix} g \begin{bmatrix} I_p & 0 \\ 0 & -I_{n-p} \end{bmatrix},$$

where also $0 \le p \le n$ and $m = n/2$ where n is even. It can be shown that a biholomorphic involution of Type I or II on $\text{GL}(n,\mathbb{C})$ is similar to one of the above examples.

Groups of matrices arise in the following three ways:

I. $\mathbf{G}(x) = \{g \in \mathrm{GL}(n, \mathbb{C}) | g = g^x\}$ where ^{x} is a Type I involutive automorphism.

II. $\mathbf{G}(+) = \{g \in \mathrm{GL}(n, \mathbb{C}) | g = g^+\}$ where $^+$ is a Type II involutive automorphism.

III. $\mathbf{G}(x, +) = \{g \in \mathrm{GL}(n, \mathbb{C}) | g = g^x = g^+\}$ where ^{x} and $^+$ are commuting involutive automorphisms such that x is Type I and $+$ is Type II.

By checking the list of classical simple Lie groups (see [Hlg]) one sees that they are (except for the addition of a determinant $= 1$ condition) all of one of the above types. The group $\mathbf{G}(x)$ with

$$g^x = \begin{bmatrix} I_p & 0 \\ 0 & -I_{n-p} \end{bmatrix} g^{*-1} \begin{bmatrix} I_p & 0 \\ 0 & -I_{n-p} \end{bmatrix}$$

is the classical group $\mathrm{U}(p, n - p)$, the main concern of Chapter 5; if $g^x = \bar{g}$, then $\mathbf{G}(x) = \mathrm{GL}(n, R)$; while if

$$g^* = \begin{bmatrix} 0 & I_m \\ -I_m & 0 \end{bmatrix} \bar{g} \begin{bmatrix} 0 & -I_m \\ I_m & 0 \end{bmatrix}$$

then $\mathbf{G}(x) = \mathrm{U}^*(2m)$.

The next theorem uses the following notation: If $\mathbf{M}$ is a full-range simply invariant subspace, $\mathbf{M}$ has a representation $\mathbf{M} = \Psi H_n^2$ for a Ψ with $\Psi^{\pm 1} \in L_{n \times n}^\infty$. If $g \to g^x$ is an involutive automorphism of Type I, let $\mathbf{M}^x = \Psi^x H_n^{2\perp}$; similarly, if $\mathbf{N} = \Psi H_n^{2\perp}$ is full-range simply $*$-invariant, we define $\mathbf{N}^x$ by $\mathbf{N}^x = \Psi^x H_n^2$. By the definition of Type I, $\mathbf{M} \to \mathbf{M}^x$ is a well defined involution on the collection of full-range simply invariant or $*$-invariant subspaces. If $g \to g^+$ is a Type II involutive automorphism, we define $[\Psi H_n^2]^+ = \Psi^+ H_n^2$ and $[\Psi H_n^{2\perp}]^+ = \Psi^+ H_n^{2\perp}$; by the defining property of Type II, this also is a well defined involution of invariant subspaces.

The following theorem summarizes the invariant subspace representations for g's from the classical Lie groups. Direct proofs of this theorem and of the next will be omitted here. Let P denote the orthogonal projection of L_n^2 onto H_n^2.

THEOREM 6.B.1. *Let* $\mathbf{M}$ *be a given full-range simply invariant subspace of* $L^2(\mathbb{C}^n)$ *and let* $g \to g^x$ *and* $g \to g^+$ *be given commuting involutive automorphisms of Type* I *and Type* II *respectively. Then* $\mathbf{M}$ *has the representation* $\mathbf{M} = [\Xi H_n^\infty]^-$ *for some matrix function* Ξ *with* $\Xi^{\pm 1} \in L_{n \times n}^2$, *such that* $\Xi P \Xi^{-1}$

defines a bounded operator on L_n^2 and

such that	if and only if
1. $\Xi(e^{it}) \in \mathbf{G}(x)$	$\mathbf{M}^x + \mathbf{M} = L_n^2$
2. $\Xi(e^{it}) \in \mathbf{G}(+)$	$\mathbf{M} = \mathbf{M}^+$
3. $\Xi(e^{it}) \in \mathbf{G}(x,+)$	$\mathbf{M}^x + \mathbf{M} = L_n^2$ and $\mathbf{M} = \mathbf{M}^+$
4. $\Xi^{x^{-1}} W \Xi = constant$, where $W = W^{x^{-1}}$ is given	$W^{-1}\mathbf{M}^x + \mathbf{M} = L_n^2$
5. $\Xi^{+^{-1}} V \Xi = constant$, where $V = V^{+^{-1}}$ is given	$\mathbf{M} = V^{-1}\mathbf{M}^+$
6. $\Xi^{x^{-1}} W \Xi$ and $\Xi^{+^{-1}} V \Xi$ are constant, where $W = W^{x^{-1}}$ and $V = V^{+^{-1}}$ are given	$W^{-m}\mathbf{M}^x + \mathbf{M} = L_n^2$, $\mathbf{M} = V^{-1}\mathbf{M}^+$, and $(V^x W V^{-1} W^{+^{-1}})\mathbf{M}^{x+} = \mathbf{M}^{x+}$
7. $\Xi^{x^{-1}} W \Xi = D$ where $W = W^{x^{-1}}$ is given, $D = D^{x^{-1}}$ is a winding matrix in a canonical form appropriate for ^{x} (Table 6.1 below)	$\{W^{-1}\mathbf{M}^x, \mathbf{M}\}$ is a Fredholm pair of subspaces of L_n^2

The reader should note that statements 1, 2, and 3 in the theorem follow from 4, 5, and 6 respectively, but with the extra content in 1, 2, and 3 that the constants can be chosen to be the identity.

Finally, we give a summary of the general factorization results arising immediately from these invariant subspace representations.

THEOREM 6.B.2. *Let W, V, and F be given functions in $L_{n\times n}^\infty$ with inverses W^{-1}, V^{-1}, and F^{-1} also in L_n^∞, let $g \to g^x$ and $g \to g^+$ be given commuting involutive automorphisms of Type I and Type II respectively, and suppose that $W = W^{x^{-1}}$ and $V = V^{+^{-1}}$. In order that F have a factorization $F = \Xi G$ with $G^{\pm 1} \in H_{n\times n}^2$ such that $G^{-1} P G$ defines a bounded operator on L_n^2 and*

such that	it is necessary and sufficient that
1. $\Xi^{x^{-1}} W \Xi$ is a contant	$T_{F^{x^{-1}}WF}$ is invertible
2. $\Xi^{+^{-1}} V \Xi$ is a constant	$(F^{+^{-1}} V F)^{\pm 1} \in H^\infty(M_n)$
3. Both $\Xi^{x^{-1}} W \Xi$ and $\Xi^{+^{-1}} V \Xi$ are constant	$T_{F^{x^{-1}}WF}$ is invertible, $(F^{+^{-1}} V F)^{\pm 1} \in H^\infty(M_n)$ and $[F^{-1}V^{-1}W^{+^{-1}}V^x W F]^{\pm 1} \in H^\infty(M_n)$
4. $\Xi^{x^{-1}} W \Xi = D$ where $D = D^{x^{-1}}$ is a winding matrix in an appropriate canonical form	$T_{F^{x^{-1}}WF}$ is Fredholm

In connection with line 7 in Theorem 6.B.1 and line 4 of Theorem 6.B.2, the following Table 6.B.1 gives the appropriate canonical form for D.

Type I involution^x

Canonical form for winding matrix $D = D^{x^{-1}}$

1. $g^x = g^{*-1}$

$$D(e^{it}) = \begin{bmatrix} e^{-ik_1 t} & & & & & & & & & \\ & e^{-ik_2 t} & & & & & & & & \\ & & \ddots & & & & & & & \\ & & & e^{-ik_r t} & & & & & & \\ & & & & \begin{bmatrix} I_{\alpha+} & e \\ e & -I_{\alpha-} \end{bmatrix} & & & & & \\ & & & & & e^{ik_r t} & & & & \\ & & & & & & \ddots & & \\ & & & & & & & e^{ik_2 t} & \\ & & & & & & & & e^{ik_1 t} \end{bmatrix}$$

where $k_1 \geq k_2 \geq \cdots \geq k_r > 0$, $p + q + 2r = n$

2. $g^x = \bar{g}$

$D(e^{it}) = \text{diag}\{e^{ik_1 t}, \ldots, e^{ik_n t}\}$

3. $g^x = \begin{bmatrix} 0 & I_m \\ I_m & 0 \end{bmatrix} \bar{g} \begin{bmatrix} 0 & -I_m \\ I_m & 0 \end{bmatrix}$

$D(e^{it}) = \begin{bmatrix} -D_1(e^{it}) & 0 \\ 0 & D_1(e^{it}) \end{bmatrix}$

where $D_1(e^{it}) = \text{diag}\{e^{ik_1 t}, \ldots, e^{ik_m t}\}$

TABLE 6.B.1

Another consequence of this study is that there are several natural interpolation problems for matrix-valued functions satisfying symmetries. The problem is: *given interpolating data, does an* $\overline{\mathbf{B}}H_{n\times n}^{\infty;l}$ *function exist which meets the interpolating constraints and which satisfies a symmetry condition?* See [**BH3**] for details. In turns out that for canonically chosen data, there is a Pick matrix whose positivity tells the full story. A new wrinkle is that in some cases merely specifying $F(z_j)\eta_j = \mu_j$ is not enough. One needs a constraint on $F'(z_j)$ of some functional of $F'(z_j)$ and this enters the Pick matrix. Table 6.B.2 indicates which symmetries can be treated, what group they correspond to, and what features of $F'(z_j)$ are critical.

Group	Interpolating function		Canonical inter-polation problem is $F(z_j)\eta_j = \mu_j$ plus	Group is conjugate to
	Constrained to satisfy	Its values are matrices in		
$U(p,q)$	None	M_{pq}	Nothing	
$O(n,\mathbb{C})$			No interp. prob.	
$O(p,q)$	$F^T F = I$	M_{pq}	$\langle F'(z_j)\eta_j, \nu_j\rangle_{\mathbb{C}^p}^R = 0$	
$U(n,n)\cap\delta_n$	$F^T = -F$	M_n	$\langle F'(z_j)\eta_j, \eta_j\rangle_{\mathbb{C}^n}^R = 0$	$O^*(2n)$
$U(n,n)\cap Sp(n,\mathbb{C})$	$F^T = F$	M_n	$\langle F'(z_j)\eta_j\eta_j\rangle_{\mathbb{C}^n} = \rho_j$	$Sp(n,R)$
$U(2p,2q)\cap\gamma_{p,q}$	$F^T J_p F = -J_q$	$M_{2p,2q}$	$\langle F'(z_j)\eta_j, J_p, \nu_j\rangle_{\mathbb{C}^{2p}} = \rho_j$	$Sp(p,q)$

TABLE 6.B.2

C. Interpolation on the boundary. We shall only give references and little explanation. There are several types of interpolation on the boundary. First they can be subdivided into two categories.

One is *infinite* and the other *finite*. A typical infinite question is: *Given* $f(e^{i\theta})$ *for* θ *on a set* E *of positive measure, does* f *continue to a function in* $\mathbf{B}H^\infty$? The questions and solutions at various levels of generality are classical. From an operator theorist's viewpoint (e.g. commutant lifting) this problem is substantially easier than Nevanlinna-Pick type interpolation (cf. [**RR2**]). For an invariant subspace treatment of the "infinite problem" see Chapter 3.D [**BH1**].

A finite problem is: *Given a finite number of* $z_k, w_k \in \mathbb{C}$ *with* $|z_k| = 1$ *and* $|w_k| \leq 1$, *find a continuous* f *in* $\mathbf{B}H^\infty$ *such that* $f(z_j) = w_j$. This turns out to be easy in the sense that a solution always exists and can in principle be constructed. Difficulties begin when an *extra* constraint is added, namely, *require that* $f'(z_j)$ *be specified*. This is the boundary analog of the Nevanlinna-Pick problem, although it is fundamentally a bit harder to analyze. Operator-theoretic approaches have serious difficulties with this approach. First of all a study set in L_n^2 of the circle obviously won't work because there is no way to discuss evaluation at a point z_j on the circle. Thus commutant lifting or [**RR2**] does not apply. Our invariant subspace signed form approach [,] can be made to work however. The point is that while the signed form [,] is canonical the space L_N^2 on which it sits is not. Indeed one can find a modifiction of L_N^2 which serves as a Krein space appropriate to our approach. This is done in [**BH5**]. An interested reader should also see [**DD**] for an extremely powerful treatment of this problem.

A cultural remark is that the difficult finite interpolation problem is basic in electric circuit theory (cf. [**YS**], [**H3, 4**], and [**DD**]).

D. Commutant lifting. Given $\mathbf{H}$ and $\tilde{\mathbf{H}}$ two Hilbert spaces and two contractions T and $\tilde{T}$ on them, $T \in \mathbf{L}(\mathbf{H})$ and $\tilde{T} \in \mathbf{L}(\tilde{\mathbf{H}})$. Let U and $\tilde{U}$ denote their minimal isometric dilations to $\mathbf{K}$ and $\tilde{\mathbf{K}}$ (see [**NF**] for discussions and definition). Consider a fixed operator A in

$$\mathbf{T}(\tilde{T}, T) \triangleq \{A \in \mathbf{L}(\mathbf{H}, \tilde{\mathbf{H}}) \colon \tilde{T}A = AT\},$$

the set of all operators which intertwines T and $\tilde{T}$. *Find the set*

$$\mathrm{CID}^l(A) = \left\{ \begin{array}{l} A_\infty \in \mathbf{L}(\mathbf{K}, \tilde{\mathbf{K}}) \colon \tilde{U}A_\infty = A_\infty U, I - A_\infty^* A_\infty \text{ has} \\ \text{at most } l \text{ negative eigenvalues and } PA = AP \end{array} \right\}$$

of all almost contractive intertwining dilations of A. Here $P \colon \mathbf{K} \to \mathbf{H}$ and $\tilde{P} \colon \tilde{\mathbf{K}} \to \tilde{\mathbf{H}}$ are orthogonal projections.

This section is not self-contained (as the reader has seen by now) and supposes some familiarity with commutant lifting theory. The Commutant Lifting Theorem is extremely powerful in that it subsumes the most general types of interpolation, H^∞ approximation, and Toeplitz Corona Theorems. For details the reader is referred to [**NF**], [**RR**], and [**S**]. Arsene, Ceausescu, and Foiaş [**ACF**] parametrize explicitly the set $\mathrm{CID}^0(A)$ of all contractive intertwining dilations. Our objective is to sketch how one obtains such a parametrization. We omit some details and refer the reader to §4 of [**BH1**] for a thorough treatment.

Begin by recalling from classical dilation theory (à la [**NF**]) that

$$\mathbf{K} = \mathbf{H} \oplus \mathbf{M}_+$$

where $\mathbf{L} = \mathrm{clos}[(U - T)\mathbf{H}]$ and $\mathbf{M}_+ = \bigoplus_0^\infty U^n \mathbf{L}$; similarly $\tilde{\mathbf{K}} = \mathbf{H} \oplus \tilde{\mathbf{M}}_+$. Now we construct the Krein space $\hat{\mathbf{K}}$ and subspace $\mathbf{M}$ which fits this problem into our approach. Let $\hat{\mathbf{K}} = \tilde{\mathbf{K}} \boxplus \mathbf{K}$ be the Krein space with inner product

$$[\tilde{k} \boxplus k, \tilde{k} \boxplus k]_{\hat{\mathbf{K}}} = [\tilde{k}, \tilde{k}]_{\tilde{\mathbf{K}}} = [k, k]_{\mathbf{K}}$$

and define a subspace $\mathbf{M}$ by

$$\mathbf{M} = \begin{bmatrix} A \\ I \end{bmatrix} \mathbf{H} \boxplus \begin{bmatrix} \tilde{\mathbf{M}}_+ \\ \mathbf{M}_+ \end{bmatrix}.$$

Here $\oplus$ denotes the sum of orthogonal spaces in the definite inner product and $\boxplus$ in the signed one. If we let $\hat{U}$ be the operator $\begin{bmatrix} \tilde{U} & 0 \\ 0 & U \end{bmatrix}$ on $\hat{\mathbf{K}}$, then $\hat{U}$ is isometric in the $[\ ,\ ]_{\hat{\mathbf{K}}}$-inner product, and since

$$\hat{U} \begin{bmatrix} A \\ I \end{bmatrix} = \begin{bmatrix} \tilde{U} & 0 \\ 0 & U \end{bmatrix} \begin{bmatrix} A \\ I \end{bmatrix} = \begin{bmatrix} \tilde{T} & 0 \\ 0 & U \end{bmatrix} \begin{bmatrix} A \\ I \end{bmatrix} + \begin{bmatrix} \tilde{U} - \tilde{T} & 0 \\ 0 & U - T \end{bmatrix} \begin{bmatrix} A \\ I \end{bmatrix}$$

$$= \begin{bmatrix} A \\ I \end{bmatrix} T + \begin{bmatrix} (\tilde{U} - \tilde{T})A \\ U - T \end{bmatrix},$$

we see that $\mathbf{M}$ is invariant for $\hat{U}$. Furthermore, $A_\infty \in \mathbf{L}(\mathbf{K}, \tilde{\mathbf{K}})$ intertwines $\tilde{U}$ and $U (\tilde{U} A_\infty = A_\infty U) \Leftrightarrow$ its graph $\begin{bmatrix} A_\infty \\ I \end{bmatrix} \mathbf{L}$ is invariant under $\hat{U}$, and $\tilde{P} A_\infty = AP \Leftrightarrow$ the graph $\begin{bmatrix} A_\infty \\ I \end{bmatrix} \mathbf{K}$ is contained in $\mathbf{M}$, and $\|A_\infty\| \leq 1 \Leftrightarrow$ the graph $\begin{bmatrix} A_\infty \\ I \end{bmatrix} \mathbf{K}$ is a negative subspace of $\hat{\mathbf{K}}$. As we have seen in Chapter 3, a $\hat{\mathbf{K}}$-maximal negative subspace is always the graph of a contraction operator. Similarly $A_\infty \in \mathrm{CID}_l(A)$ if there is some U-invariant subspace $\psi\mathbf{K}$ of codimension at most l as a subspace of $\mathbf{K}$ such that $A_\infty : \psi\mathbf{K} \to \hat{\mathbf{K}}$ satisfies $\|A_\infty\| \leq 1$, $A_\infty(U|\psi\mathbf{K}) = \tilde{U} A_\infty$, and $\tilde{P} A_\infty = AP|\psi\mathbf{K}$. (Thus $\tilde{P} A_\infty|\mathbf{H}$ agrees with A on a T-invariant subspace of codimension at most l.) Using the same analysis as usual, we obtain

PROPOSITION 6.D.1. *The angle operator-graph correspondence establishes a one-to-one correspondence between the set $\mathrm{CID}^l(A)$ and $\hat{U}$-invariant negative subspaces of $\mathbf{M}$ of codimension at most l in a $\hat{\mathbf{K}}$-maximal negative subspace of $\hat{\mathbf{K}}$.*

Thus by Fact 3.C.5, a necessary condition that the set $\mathrm{CID}^l(A)$ be nonempty is that the negative signature of $\mathbf{M}'$ be at most l, and conversely, once we establish that there exist $\mathbf{M}$-maximal negative subspaces of $\mathbf{M}$ which are also $\hat{U}$-invariant, we shall see that this condition is sufficient as well. Now it is easily checked that

$$\mathbf{M}' = \begin{bmatrix} I \\ A^* \end{bmatrix} \tilde{\mathbf{H}},$$

and thus the negative signature of $\mathbf{M}'$ is the dimension of the negative spectral subspace for the selfadjoint operator $I - AA^*$ on $\tilde{\mathbf{H}}$.

Now we apply the invariant subspace representation, Corollary 5.1, to $\mathbf{M}$. Actually we need to generalize from $M_{e^{i\theta}}$ to an "abstract shift" $\hat{U}$. Also possibly $\hat{U}|_\mathbf{M}$ is *not* simply invariant (e.g. take $A = 0$ and let T be any c.n.u. contraction which is not C_0. Such a generalization to include a $[\ ,\ ]_{\hat{\mathbf{K}}}$ Wold decomposition is straightforward and is left as a very long exercise (cf. Theorem 4.1 [**BH1**] for the solution). As a consequence we obtain the following generalization of the existence of the Sz. Nagy–Foiaş lifting theorem (which corresponds to the $l = 0$ case).

THEOREM 6.D.2. *The set $\mathrm{CID}^l(A)$ is nonempty if and only if the negative spectral subspace of $I - AA^*$ on $\mathbf{H}$ has dimension at most l.*

E. F. and M. Riesz theorem. *If ν is a Borel measure on the circle such that $\int e^{in\theta} d\nu = 0$ for all $n > 0$, then ν is absolutely continuous w.r.t. Lebesgue measure and there exists f in H^1 such that $d\nu = f\, d\theta$.*

This is one of the traditional applications of the Beurling theorem and we include it here so that the reader can see how this sort of argument goes. Our presentation follows R. G. Douglas [**D**] closely.

PROOF. If μ denotes the total variation of ν, then there exists a Borel function ψ such that $d\nu = \psi\, d\mu$ and $|\psi| = 1$ almost everywhere w.r.t. μ. Let $\mathbf{M}$ denote the closed subspace of $L^2(\mu)$ generated by $\{e^{in\theta} : n > 0\}$. If $\mathbf{M}$ is simply invariant, then $\mathbf{M} = \phi H^2$ for a function ϕ with $|\phi|^2 \, d\mu = d\theta/2\pi$ by the classical Beurling theorem. Since $e^{i\theta}$ is in $\mathbf{M}$, it follows that there is a g in H^2 such that $e^{i\theta} = \phi g$ almost everywhere w.r.t. μ. Consequently $\phi \neq 0$ a.e. w.r.t. μ and so μ is absolutely continuous w.r.t. Lebesgue measure; that is, $d\nu = f \, d\theta$ for an f in L^1. The hypothesis of the theorem implies that the negative Fourier coefficients of f vanish, so f is in H^1.

If $\mathbf{M}$ is not simply invariant, then it is easy to show (cf. Theorem 6.12 [**D**]) that $\mathbf{M} = \mathbf{M}_1 \oplus \mathbf{M}_2$ where $\mathbf{M}_1$ is simply invariant and $\mathbf{M}_2$ is doubly invariant (that is, $\mathbf{M}_2$ is $e^{\pm i\theta}$ invariant). Doubly invariant subspaces of $L^2(\mu)$ are all of the form $L^2(E; \mu)$ for a Borel subset E of the unit circle; so $\mathbf{M}_2 = L^2(E; \mu)$. Now

$$(e^{in\theta}, \overline{\psi}) = \int_{\mathbb{T}} e^{in\theta} \psi \, d\mu = \int_{\mathbb{T}} e^{in\theta} \, d\nu = 0$$

and hence $\overline{\psi}$ is orthogonal to $\mathbf{M}$ in $L^2(\mu)$. Moreover,

$$\mu(E) = \int_{\mathbb{T}} |\psi|^2 I_E \, d\mu = (\overline{\psi}, \overline{\psi} I_E) = 0,$$

since $\overline{\psi} I_E$ is in $\mathbf{M}_2$ and $\overline{\psi}$ is orthogonal to $\mathbf{M}$. Here I_E is the characteristic function of E. Therefore $\mathbf{M}_2 = \{0\}$ and so $\mathbf{M} = \mathbf{M}_1$ is indeed simply invariant. Q.E.D.

F. Completely integrable systems. Completely integrable systems of ordinary differential equations (such as those arising from the Toda lattice) and of partial differential equations (such as the Korteweg-de Vries) have a magical connection with Wiener-Hopf factorization. Here we shall present a few examples with emphasis on the role that Wiener-Hopf factorization plays. Time is too short and knowledge too limited to describe anything close to the full picture.

The Toda lattice example here was shown to me by Nolan Wallach and one can see his paper [**GW**] for history and motivation. The account here of the Zachorov-Shabat approach to the modified KdV equations comes from lectures by John Palmer.

Without motivation we assert that one type of "Toda lattice" gives rise to the following system of differential equations. Let $X(s)$ be a matrix-valued function. Let $X_+(s)$ be its upper triangular part, and $X_-(s)$ is its strictly lower triangular part. That is, $X(s) = X_+(s) + X_-(s)$. We want to *find solutions to*

(6.2) $$\frac{dX(s)}{ds} = [X_+(s), X_-(s)].$$

SOLUTION: Fix a matrix X^0 and let $L(s)U(s) = e^{sX^0}$ be a lower-upper triangular factorization of e^{sX^0}. Here $L(s)$ is normalized to have all ones on the diagonal. *We assume that* such factors exist (they must at least for small s). Define a matrix function

$$X(s) = U(s)X^0 U(s)^{-1}.$$

Since $LUX^0 = X^0 LU$, we get $X(s) = L(s)^{-1} X^0 L(s)$.

PROPOSITION 6.F.1. $X(s)$ is a solution to (6.2) with $X(0) = X^0$.

PROOF. By definition of L and U

$$L'U + LU' = (LU)' = (e^{sX^0})' = X^0 e^{sX^0} = LUX^0,$$

so

$$L^{-1}L' + U'U^{-1} = UX^0 U^{-1} = X(s).$$

The first term is strict lower triangular, the second upper triangular, so

$$X_+(s) = U'U^{-1} \quad \text{and} \quad X_-(s) = L^{-1}L'.$$

This allows us to check

$$\frac{dX(s)}{ds} \overset{?}{=} [X_+(s), X_-(s)] = [X_+(s), X(s)],$$

$$U'X^0 U^{-1} - UX^0 U^{-1} U'U^{-1} \overset{?}{=} [U'U^{-1}, UX^0 U^{-1}]$$
$$= U'X^0 U^{-1} - UX^0 U^{-1} U'U^{-1}. \quad \text{Q.E.D.}$$

Thus we have built solutions by an upper-triangular lower-triangular analog of a Wiener-Hopf argument.

The next example uses a more traditional Wiener-Hopf argument to produce solutions to the modified KdV equations.

One way look at the subject is elegant though a bit roundabout. Suppose that we want to see if the first order system of equations

$$(6.3) \qquad \frac{\partial}{\partial t_j} \psi(z; t_1, t_2, \ldots) = V_j(z; t_1, t_2, \ldots) \psi(z; t_1, t_2, \ldots)$$

has a simultaneous solution ψ. At its most general level, this business is formal, and traditionally, practitioners allow infinitely many variables. In the examples we present here the most we shall ever use is two variables t_1 and t_2. Here V_j is a $N \times N$ matrix-valued function of the t's and of z a complex variable, while ψ is an R^N-valued function.

To have a simultaneous solution the cross partials must be equal; that is,

$$(6.4) \qquad \begin{aligned} \frac{\partial}{\partial t_i} \frac{\partial \psi}{\partial t_j} &= \frac{\partial}{\partial t_i} V_j \psi = \frac{\partial V_j}{\partial t_i} \psi + V_j \frac{\partial \psi}{\partial t_i}, \\ \frac{\partial}{\partial t_j} \frac{\partial \psi}{\partial t_i} &= \frac{\partial}{\partial t_j} V_i \psi = \frac{\partial V_i}{\partial t_j} \psi + V_i \frac{\partial \psi}{\partial t_j}, \end{aligned}$$

so

$$\left\{ \frac{\partial V_j}{\partial t_i} - \frac{\partial V_i}{\partial t_j} + V_j V_i - V_i V_j \right\} \psi = 0.$$

If there are enough solutions ψ so that the simultaneous equations (6.3) have a fundamental solution, then

$$(6.5) \qquad \frac{\partial V_j}{\partial t_i} - \frac{\partial V_i}{\partial t_j} + [V_j, V_i] = 0.$$

Indeed (6.5) is the basic set of compatibility conditions for (6.3).

Wiener-Hopf factorization enters because we want to find V_j's satisfying (6.5) which have "simple" dependence on z. In particular in this discussion we wish to solve

MAIN PROBLEM: Fix an $N \times N$ matrix F satisfying $\operatorname{tr} F^{\mu_j} = 0$. Find $N \times N$ matrix-valued functions $V_j(z_j; t_1, t_2, \ldots)$ such that

(i)   the functions satisfy equations (6.5),

(ii)

$$V_j(z; t) = \sum_{k=0}^{\mu_j} v_{jk}(t) z^k$$

is a matrix polynomial with highest order terms $= F^{\mu_j} z^{\mu_j}$, and

(iii) $\operatorname{tr} V_j(z; t) \equiv 0$ (provided $\operatorname{tr} F^{\mu_j} = 0$).

Here $t = (t_1, t_2, \ldots)$.

EXAMPLE 1. Take $\mu = 2$, take $t = (t_1, t_2)$, and take $F = \begin{pmatrix} 0 & 1 \\ 1 & 0 \end{pmatrix}$.

$$V_1 = \begin{pmatrix} v(t) & 0 \\ 0 & -v(t) \end{pmatrix} + \begin{pmatrix} 0 & 1 \\ 1 & 0 \end{pmatrix} z;$$

$$V_2 = \begin{pmatrix} a(t) & 0 \\ 0 & -a(t) \end{pmatrix} + \begin{pmatrix} 0 & b(t) \\ c(t) & 0 \end{pmatrix} z + \begin{pmatrix} d(t) & 0 \\ 0 & -d(t) \end{pmatrix} z^2 + \begin{pmatrix} 0 & 1 \\ 1 & 0 \end{pmatrix} z^3.$$

Then one can by direct computation check that V_1 and V_2 satisfy (6.5) if and only if

$$\frac{dv}{dt_2} = \frac{\partial^3 v}{\partial t_1^3} + 6v \frac{\partial v}{\partial t_1}.$$

This is the *modified* KdV equation. You see that the mKdV equation arises by requiring that we solve (6.5) subject to side conditions (ii) and (iii). Indeed other "completely integrable" equations arise in a similar way.

Consequently generating V_j's which solve the *Main Problem* is a worthy cause.

EXAMPLE 2. Take $V_j^0 \triangleq F^j z^j$. Then the system (6.3) has a fundamental solution ψ_0 given formally by

$$\psi_0(z, t) = \exp \sum_{k=0} t_k F^k z^k.$$

For a finite system, say $j = 1, \ldots, M$, the sum stops at M. Obviously since (6.3) has a solution, the V_j^0's are solutions to (6.5). Thus V_j^0 is a solution to (i) and (ii) of the Main Problem. To get the trace $= 0$ condition, we only permit F^j which satisfy $\operatorname{tr} F^j = 0$.

The observation we describe here is due to Zachorov and Shabat and is that the solution given in Example 2 generates many solutions to the Main Problem.

RECIPE 6.F.2. *The recipe for solving the Main Problem is:*

(1) *Pick* $g\colon \mathbf{T} \to \mathrm{SL}(N, \mathbb{C})$.

(2) *Set* $\phi(e^{i\theta}; t) = \psi_0(e^{i\theta}; t)g(e^{i\theta})\psi_0(e^{i\theta}; t)^{-1}$.

(3) *Factor* $\phi(e^{i\theta}; t) = \phi_-(e^{i\theta}; t)\phi_+(e^{i\theta}; t)$ *where*

$$\phi_-^{\pm 1} \text{ is holomorphic in } \|z\| \geq 1,$$

$$\phi_+^{\pm 1} \text{ is holomorphic in } \|z\| \leq 1,$$

normalized so that $\phi_-(t, \infty) = \mathrm{id}$.

(4) *Set* $\psi = \phi_-^{-1}\psi_0$ *and*

$$V_j = -\phi_-^{-1}\frac{\partial \phi}{\partial t_j} + \phi_-^{-1}V_j^0 \phi_-.$$

Then V_j *solves the Main Problem and* ψ *solves the corresponding system* (6.5). *Note: A* t *at which* ϕ *does not factor yields a singularity of* V_j.

SKETCH OF PROOF. It is easy to generate functions V_j so that the system (6.3) has a fundamental solution. In fact let ϕ_- be any invertible matrix-valued function. Then V_j given by (4) gives a system (6.3) with solution $\psi = \phi_-^{-1}\psi_0$. The verification is

(6.6)
$$-\left(\frac{\partial}{\partial t_j} - V_j\right) = \frac{\partial}{\partial t_j} + \phi_-^{-1}\frac{\partial \phi}{\partial t_j} - \phi_-^{-1}V_j^0 \phi_-$$
$$= \phi_-^{-1}\left(\frac{\partial}{\partial t_j} - V_j^0\right)\phi_-,$$

so

$$\left(\frac{\partial}{\partial t_j} - V_j\right)\phi_-^{-1}\psi_0 = \phi_-^{-1}\left(\frac{d}{dt_j} - V_j^0\right)\psi_0 = 0.$$

The Wiener-Hopf definition of ϕ_0 in (3) has the effect of forcing V_j to be a polynomial with the same order as V_j^0. This clever observation is based on the fact that ϕ was selected so that

$$\left(\frac{\partial}{\partial t_j} - V_j^0\right)M_\phi = M_\phi\left(\frac{\partial}{\partial t_j} - V_j^0\right).$$

Consequently,

$$\phi_-^{-1}\left(\frac{\partial}{\partial t_j} - V_j^0\right)\phi_- = \phi_+\left(\frac{\partial}{\partial t_j} - V_j^0\right)\phi_+^{-1},$$

$$\frac{\partial}{\partial t_j} - V_j = \frac{\partial}{\partial t_j} + \phi_+\frac{\partial \phi_+^{-1}}{\partial t_j} - \phi_+ V_j^0 \phi_+^{-1},$$

so

$$V_j = -\phi_-^{-1}\frac{\partial \phi_-}{\partial t_j} \quad + \phi_-^{-1}V_j^0\phi_- \quad = -\phi_+\frac{\partial \phi_+^{-1}}{\partial t_j} \quad + \phi_+ V_j^0\phi_+^{-1},$$

| $\uparrow$ | $\uparrow$ | $\uparrow$ | $\uparrow$ |
| analytic outside disk | analytic outside except for jth order pole at ∞ | analytic inside | analytic inside |

which from our analysis of poles is analytic inside the disk except for exactly j poles at ∞. Thus V_j is a polynomial of order j. The normalization $\phi_-(\infty) =$ identity implies that the highest order term of V_j equals the highest order term of $V_j^0 = F^j z^j$.

This leaves only (iii) to check. It is straightforward, though there are more things to check than before. To get that each $V_j(e^{i\theta}; t)$ has trace $= 0$, start with the fact that V_j^0 has trace $= 0$, i.e., is in the Lie algebra of $SL(N, \mathbb{C})$. From (6.6) note that V_j satisfies

$$\frac{\partial}{\partial t_j} - V_j = \phi_-^{-1}\left(\frac{\partial}{\partial t_j} - V_j^0\right)\phi_-$$

and so, trace $V_j = 0$ provided $\phi_- \in SL(N, \mathbb{C})$.

To prove that $\phi_- \in SL(N, \mathbb{C})$ we first note that if $\phi \in SL(N, \mathbb{C})$, then its Wiener-Hopf factors satisfy $1 = \det\phi = \det\phi_+ \det\phi_-$. Since $\det\phi_- = 1/\det\phi_+ \in H^\infty$ and $\det\phi_- \in H^\infty +$ constants, we get $\det\phi_- =$ constant which we normalize to 1. Consequently $\phi \in SL(N, \mathbb{C})$ implies $\phi_- \in SL(N, \mathbb{C})$.

Now $\phi = \psi_0 g \psi_0^{-1}$. Since $g(e^{i\theta})$ is in $SL(N, \mathbb{C})$, we need only show that ψ_0 is in $SL(N, \mathbb{C})$. However, it is, because $\partial\psi_0/\partial t_j = V_j^0\psi_0$ where V_j^0 is in the Lie algebra of $SL(N, \mathbb{C})$.

G. Segal and G. Wilson [**SW**] gave an "invariant subspace" viewpoint to the recipe. It is most elegant and, since it is of a Grassmannian nature, the occasional singularities of solutions are easy to visualize.

The setting is L_N^2. The geometric picture is simple to the point of being banal. Then one must interpret it algebraically which is indeed more work.

Roughly, one considers a full range simply invariant subspace $\mathbf{M}$. Define $\mathbf{M}_t = \{\psi_0(e^{i\theta}; t)f(e^{i\theta}): f \in \mathbf{M}\}$; the many parameter group of multiplication operators $\psi_0(\ ;t)$ induces a flow on the invariant subspaces. This *is* the geometric picture.

To formulate this algebraically, pair $\mathbf{M}_t$ with $\mathbf{M}^x \triangleq \overline{H}_N^2$. Then represent this using Theorem 4.1 as $\mathbf{M}_t = \text{clos}\,\phi_-(\ ;t)H_N^2$ and $\overline{H}_N^2 = \text{clos}\,\phi_-(\ ;t)\overline{H}_N^2$, provided $\mathbf{M}_t + \overline{H}_N^2 = L_N^2$ and $\mathbf{M}_t \cap \overline{H}_N^2 = \{0\}$. The second equality is equivalent to ϕ_- being *-outer.

PROPOSITION 6.F.3. *Represent* $\mathbf{M}$ *as* $\mathbf{M} = gH_N^2$. *If* ϕ_- *is normalized correctly, then it is indeed the negative Wiener-Hopf part of* $\psi_0 g \psi_0^{-1}$. *Thus the* $\mathbf{M}_t$ *in a sense parametrizes the solutions to the Main Problem.*

PROOF. Since $\psi_0^{\pm 1}(\ ;t) \in H_{N \times N}^\infty$, we have

$$\mathbf{M}_t = \psi_0(,t)gH_N^2 = \psi_0(\ ;t)g\psi_0(\ ;t)^{-1}H_N^2.$$

Consequently, $\phi_- H_N^2 = \psi_0 g \psi_0^{-1} H_N^2$; so $\phi_- k = \psi_0 g \psi_0^{-1}$ where k is a constant $N \times N$ matrix. The normalization issue is one of selecting k correctly. To do so involves keeping track of additional subspaces. Namely, an invariant codimension one subspace $\mathbf{M}'$ of $\mathbf{M}$, an invariant codimension one subspace $\mathbf{M}^2$ of $\mathbf{M}^1$, etc., down to $\mathbf{M}^{N-1}$. While the matter is straightforward we shall not pursue it here.

7. Matrix Analogs and Generalizations

The workhorse of previous chapters was a representation for invariant subspaces. This chapter presents a refinement in which we represent a nested family of invariant subspaces (see Figure 7.1).

While an operator theorist typically calls such a family of subspaces a nest, a group representer typically calls it a flag. Indeed Graeme Segal and George Wilson introduced and studied a very special flag of invariant subspaces called a *periodic flag*. The manifold of these, denoted $FL^{(k;n)}$, is the set of all sequences $\{\mathbf{M}_j\}_{j=1}^n$ of subspaces of L_k^2 such that

(i)   Each $\mathbf{M}_j$ is full range and simply invariant for the shift operator of multiplication by z (FRSI).

(ii) $\mathbf{M}_{j+1} \subset \mathbf{M}_j$ for each j.

(iii) $z\,\mathbf{M}_1 \subset \mathbf{M}_n$.

Segal and Wilson are interested in the manifold structure of $FL^{(k;n)}$, studying the orbits of the group of *-outer functions on $FL^{(k;n)}$, and how this induces "a stratification" and "cellular subdivision" of $FL^{(k;n)}$. These results can be seen as generalizations of classical results about the standard flag of all chains of subspaces $\mathbb{C}^n = \mathbf{M}_1 \supset \mathbf{M}_2 \supset \cdots \supset \mathbf{M}_n \supset (0)$ $(\dim \mathbf{M}_j = n + 1 - j)$ in n-dimensional space. The ultimate purpose of the machine is to study the KdV equation and the theory of Miwa, Jimbo, and Sato.

In this chapter we indicate how many of these ideas can be put into the framework of a more general $(\mathbf{M}^\times, \mathbf{M})$-theorem. A general theory emerges which includes, as special cases, the Ball-Helton $(\mathbf{M}^\times, \mathbf{M})$-theorem with all its corollaries and applications discussed in the previous chapters on the one hand, and the Ball-Gohberg $(\mathbf{M}^\times, \mathbf{M})$-theorem for finite matrices (see [**BG1**–**BG4**]). This last theory includes as applications LU and QR factorization, an alternate approach to the Arveson distance formula for finite nest algebras as well as to the positive-definite completion problem for certain special cases discussed by Charlie Johnson. As applications, we provide another treatment of a sensitivity minimization problem studied by Feintuch and Tannenbaum [**FeTa**].

73

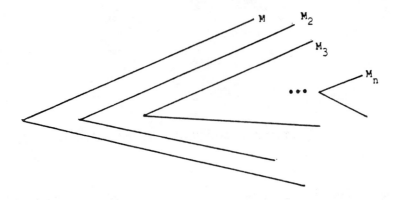

FIGURE 7.1.

A. The $(\mathbf{M}^\times, \mathbf{M})$-theorem for flags. We begin with an example of a periodic flag in H_k^p. Let $\mathbb{C}^k = F_1 \supset F_2 \supset \cdots \supset F_n \supset 0$ be a conventional flag consisting of n subspaces of $\mathbb{C}^k$ and set

(7.1)
$$\mathbf{M}_1^F = H_k^p = F_1 + e^{i\theta} H_k^p,$$
$$\mathbf{M}_2^F = F_2 + e^{i\theta} H_k^p,$$
$$\vdots$$
$$\mathbf{M}_n^F = F_n + e^{i\theta} H_k^p.$$

Call a flag of this form an L^p-*forward-reference flag.* Clearly these spaces are simply invariant and satisfy the periodicity condition $e^{i\theta} \mathbf{M}_1 \subset \mathbf{M}_n$. The main forthcoming point is that *any periodic flag* $\mathbf{M}_1 \supset \mathbf{M}_2 \supset \cdots \supset \mathbf{M}_n$ *in* L_k^2 *can be represented as multiplication by a* $g \in L_{k \times k}^2$ *acting on an* L^∞-*reference flag* $\mathbf{M}^F$ *as in* (7.1). As before this representation is highly nonunique; one needs some $\mathbf{M}^\times$ type of information to determine g uniquely.

Naturally there is a dual analysis for backward periodic flags. By a *backward periodic flag* we mean a collection of subspaces $\{\mathbf{M}_j^\times : 1 \leq j \leq n\}$ of L_k^2 such that

(i)   Each $\mathbf{M}_j^\times$ is FRSI with respect to the backward shift multiplication by z^{-1}.

(ii) $\mathbf{M}_{j+1}^\times \supset \mathbf{M}_j^\times$ for each j.

(iii) $z\,\mathbf{M}_1^\times \supset \mathbf{M}_n^\times$.

The reference example of a backward periodic flag is based on the orthogonal complement B of a flag in $\mathbb{C}^k$; thus $B = \{B_1, \ldots, B_n\}$ where $(0) = B_1 \subset B_2 \subset \cdots \subset B_n \subset \mathbb{C}^k$. The reference example then is to take

$$B_{\mathbf{M}_1^\times} = e^{-i\theta}\overline{H_k^p} = B_1 + e^{-i\theta}\overline{H_k^p},$$

$$B_{\mathbf{M}_2^\times} = B_2 + e^{-i\theta}\overline{H_k^p},$$

$$\vdots$$

$$B_{\mathbf{M}_n^\times} = B_n + e^{-i\theta}\overline{H_k^p}.$$

Call this an L^p-*backward-reference flag*.

Our main theorem and tool in this section is

THEOREM 7.1. *Suppose* $\mathbf{M}_1 \supset \mathbf{M}_2 \supset \cdots \supset \mathbf{M}_n$ *and* $\mathbf{M}_1^\times \subset \mathbf{M}_2^\times \subset \cdots \subset \mathbf{M}_n^\times$ *are forward periodic and backward periodic flags such that*

$$\mathbf{M}_1 + \mathbf{M}_1^\times = L^{k^2} \quad \text{and} \quad \mathbf{M}_1 \cap \mathbf{M}_1^\times = \{0\}.$$

Then there are two L^∞-*reference flags* $\mathbf{M}^F$ *and* $^B\mathbf{M}^\times$ *and* f *in* $L^2_{k \times k}$ *such that*

$$\mathbf{M} = \operatorname{clos} g\,\mathbf{M}^F \quad \text{and} \quad \mathbf{M}^\times = \operatorname{clos} g\,{}^B\mathbf{M}^\times.$$

Once B and F are chosen, the function g is unique up to an invertible constant right factor g_0 which respects the flags B and F on the constants

$$g_0 F_j = F_j, \quad g_0 B_j = B_j, \qquad j = 1, 2, \ldots, n.$$

Moreover, if

$$\mathbf{M}_j + \mathbf{M}_j^\times = L_k^2 \quad \text{and} \quad \mathbf{M}_j \cap \mathbf{M}_j^\times = \{0\}$$

for $1 \le j \le n$, then one can take $B_j = F_j^\perp$ for $1 \le j \le n$.

For example if $n = k$ and all "gaps" $\mathbf{M}_j/\mathbf{M}_{j+1}$ and $\mathbf{M}_{j+1}^\times/\mathbf{M}_j^\times$ have dimension one, then g_0 is diagonal (in an appropriate basis). This illustrates the main point of flags. They allow one to control the 0th Fourier coefficient of matrix-valued functions which appear in Wiener-Hopf factorization, Corona theorems, etc.

To prove the theorem and frequently to apply it, we have found a slick way of representing periodic flags most useful. It is the one used by Ball and Gohberg, extended to our setting.

Given a periodic flag $\mathbf{M}_1 \supset \mathbf{M}_2 \supset \cdots \supset \mathbf{M}_n$, introduce a subspace $\underline{\mathbf{M}} \subset L^2_{k \times n}$,

$$\underline{\mathbf{M}} = [\mathbf{M}_1, \mathbf{M}_2, \ldots, \mathbf{M}_n].$$

Thus $\underline{\mathbf{M}}$ consists of all $L^2_{k \times n}$-functions such that the jth column is an element of the subspace $\mathbf{M}_j \subset L_k^2$. Conversely, given any subspace $\underline{\mathbf{M}} \subset L^2_{k \times n}$ we would like to characterize when it arises in this way from some periodic flag $\{\mathbf{M}_j \colon 1 \le j \le n\}$. The answer is in terms of the subspace $H^\infty_{0\Omega_L}$ consisting of those functions L in $H^\infty_{n \times n}$ such that $L(0)$ is in the algebra Ω_L of lower triangular $n \times n$ matrices. Note that $H^\infty_{0\Omega_L}$ (and more generally $L^\infty_{n \times n}$) acts on $L^2_{k \times n}$ by right multiplication ($F \in L^2_{k \times n}$, $G \in L^\infty_{n \times n} \Rightarrow FG \times L^2_{k \times n}$); that is, $L^2_{k \times n}$ is a (right) module over $L^\infty_{n \times n}$. Thus it makes sense to consider $H^\infty_{0\Omega_L}$-submodules of

$L^2_{k \times n}$, i.e., subspaces $\underline{\mathbf{M}}$ of $L^2_{k \times n}$ such that $F \in \underline{\mathbf{M}}$, $L \in H^\infty_{0\Omega_L} \Rightarrow FL \in \underline{\mathbf{M}}$. To stay in the spirit of the previous chapters, let us refer to such subspaces as $H^\infty_{0\Omega_L}$-*invariant*. We say that an $H^\infty_{0\Omega_L}$-invariant subspace is $H^\infty_{0\Omega_L}$-*simply invariant* if also

$$\bigcap_{L \in H^\infty_{0\Omega_L}} \{FL \colon F \in \underline{\mathbf{M}}\} = (0)$$

and is $H^\infty_{0\Omega_L}$-*full range* if

$$\bigcup_{L \in L^\infty_{n \times n}} \{FL \colon F \in \underline{\mathbf{M}}\}$$

is dense in $L^2_{k \times n}$. Let us say that a subspace which is both $H^\infty_{0\Omega_L}$-simply invariant and $H^\infty_{0\Omega_L}$-full range is *block invariant*. Similarly a subspace which is both simply invariant and full range with respect to $\overline{H^\infty_{0\Omega_u}}$ we call *-*block invariant*.

By $\overline{H^\infty_{0\Omega_u}}$ we mean the set of all $F \in L^\infty_{k \times n}$ such that $\overline{F} \in H^\infty_{n \times n}$ and $F(0)$ belongs to the algebra Ω_u of $n \times n$ upper triangular matrices; here $\overline{F}$ denotes the complex conjugate (elementwise and pointwise) of the matrix function F. The characterization of periodic flags is now quite simple and suggestive. We do not dwell on the proof here.

PROPOSITION 7.2. *A subspace* $\underline{\mathbf{M}} \subset L^2_{k \times n}$ *is of the form*

$$\underline{\mathbf{M}} = [\mathbf{M}_1, \mathbf{M}_2, \ldots, \mathbf{M}_n]$$

for some periodic flag $\{\mathbf{M}_j \colon 1 \leq j \leq n\}$ *in* $FL^{(k;n)}$ *if and only if* $\mathbf{M}$ *is block invariant.*

The dual of Proposition 7.2 is as follows:

PROPOSITION 7.3. *A subspace* $\mathbf{M}^\times \subset L^2_{k \times n}$ *has the form*

$$\mathbf{M}^\times = [\mathbf{M}^\times_1, \mathbf{M}^\times_2, \ldots, \mathbf{M}^\times_n]$$

for a backwards periodic flag $\{\mathbf{M}^\times_j \colon 1 \leq j \leq n\}$ *if and only if* $\mathbf{M}^\times$ *is* *-*block invariant.*

We now look at some examples of block invariant subspaces of L^2_n. One is the subspace $\underline{\mathbf{M}} = H^2_{0\Omega_L}$, the subspace of $H^2_{n \times n}$ functions F such that $F(0) \in \Omega_L$. In this case $\dim \mathbf{M}_j / \mathbf{M}_{j+1} = 1$; this is the model for the case discussed after Theorem 7.1. To get more general multiplicity, let $\underline{\nu} = \{\nu_1, \ldots, \nu_n\}$ be any n-tuple of nonnegative integers and set $N(\underline{\nu}) = \nu_1 + \cdots + \nu_n$. We let $\Omega(\underline{\nu})$ be the set of $N(\underline{\nu}) \times n$ matrices, but written as $n \times n$ block matrices $F = [F_{ij}]_{1 \leq i,j \leq n}$ where the size of the (i,j) block F_{ij} is $\nu_i \times 1$. We let $\Omega_L(\underline{\nu})$ denote all such block matrices $F = [F_{ij}]_{1 \leq i,j \leq n}$ which are block lower triangular ($F_{ij} = 0$ if $i < j$). The set of block upper triangular matrices $\Omega_u(\underline{\nu})$ is defined similarly. Finally $H^2_{0\Omega_L(\underline{\nu})}$ denotes the set of $H^2_{N(\underline{\nu}) \times n}$ matrices F such that $F(0) \in \Omega_L(\underline{\nu})$. Then $H^2_{0\Omega_L(\underline{\nu})}$ is block invariant. In this case $\dim \mathbf{M}_j / \mathbf{M}_{j+1} = \nu_j$ for $1 \leq j \leq n$. Examples of *-block invariant subspaces are had by taking orthogonal complements of block invariant subspaces.

THEOREM 7.4. *Suppose* $\underline{\mathbf{M}}$ *is block invariant and* $\underline{\mathbf{M}}^\times$ *is an* *-*block invariant subspace of* $L^2_{k\times n}$, *such that*

$$L^2_{k\times n} = \underline{\mathbf{M}}^\times \dotplus \underline{\mathbf{M}}.$$

Set $\nu_j = \dim \mathbf{M}_j/\mathbf{M}_{j+1}$ *for* $1 \le j \le n$. $(\mathbf{M}_{n+1} = z\,\mathbf{M}_1$, *and* $\mathbf{M}_j$ *is the set of all vector functions occurring in the jth column of* $\mathbf{M}$.) *Then* $N(\underline{\nu}) = k$ *and there is a matrix function* $g \in L^2_{k\times k}$ *such that*

$$\underline{\mathbf{M}} = \operatorname{clos} g H^\infty_{0\Omega_L} \quad and \quad \underline{\mathbf{M}}^\times = \operatorname{clos} g\overline{H^\infty_{0\Omega_u}}.$$

(*Here the closures are in* $L^2_{k\times n}$.) *The matrix function* g *is uniquely determined up to an invertible* $\underline{\nu} \times \underline{\nu}$ *block diagonal right factor.*

This is only the canonical case of this generalized $(\mathbf{M}^\times, \mathbf{M})$-theorem. Other representations can be classified corresponding to weakening the direct sum condition $L^2_{k\times n} = \underline{\mathbf{M}}^\times \dotplus \underline{\mathbf{M}}$.

We remark that the case $n = 1$ corresponds exactly to the $(\mathbf{M}^\times, \mathbf{M})$-theorem from [**BH2**], while the case where $\mathbf{M}_1 = H^2_k$ and $\mathbf{M}^\times_1 = H^{2\perp}_k$ corresponds to the $(\mathbf{M}^\times, \mathbf{M})$-theorem for finite matrices from [**BG1**]. With various assumptions on $\mathbf{M}^\times_j$ and $\mathbf{M}_j$, one can guarantee that g is in some class of matrix functions (e.g., polynomial, rational, real analytic, or smooth); this is done in [**SW2**]. Then one can substitute $H^2_{0\Omega_L}$ for $H^\infty_{0\Omega_L}$ and dispense with clos in the representation above for $\mathbf{M}_j$.

Other theorems are obtained as corollaries by making special choices of $\mathbf{M}^\times$ in terms of $\underline{\mathbf{M}}$. A natural choice is to take $\underline{\mathbf{M}}^\times = \underline{\mathbf{M}}^{\perp J}$ for a $k \times k$ Hermitian matrix J. Here $\underline{\mathbf{M}}^{\perp J}$ is the orthogonal complement of $\mathbf{M}$ in the J-inner product on $L^2_{k\times n}$ given by

$$[F, G]_J = \frac{1}{2\pi} \int_0^{2\pi} \operatorname{tr}(G^*(e^{i\theta}) J F(e^{i\theta}))\, d\theta.$$

With this choice of $\mathbf{M}^\times$, Theorem 7.4 can be reorganized to give the following result.

THEOREM 7.5. *Let* J *be an invertible* $k \times k$ *Hermitian matrix and suppose* $\underline{\mathbf{M}}$ *is a block invariant subspace such that*

$$L^2_{k\times n} = \underline{\mathbf{M}}^{\perp J} \dotplus \underline{\mathbf{M}}.$$

Let ν^+_j *be the dimension of a maximal positive subspace of* $\mathbf{M}_j/\mathbf{M}_{j+1}$, *and let* ν^-_j *be the dimension of a maximal negative subspace of* $\mathbf{M}_j/\mathbf{M}_{j+1}$ *in the* J-inner product. *Then there is a matrix function* $g \in L^2_{k\times k}$ *such that*

(i)
$$\underline{\mathbf{M}} = \operatorname{clos} g \begin{bmatrix} H^\infty_{0\Omega_L}(\underline{\nu}^+) \\ H^\infty_{0\Omega_L}(\underline{\nu}^-) \end{bmatrix}$$

and

(ii)
$$g(e^{i\theta})^* J g(e^{i\theta}) = \begin{bmatrix} I_{N(\underline{\nu}^+)} & 0 \\ 0 & -I_{N(\underline{\nu}^-)} \end{bmatrix}.$$

Some special cases of Theorem 7.5 are worth noting explicitly. If J is the identity matrix I_k, then the $\mathbf{M}^{\perp J}$ is the Hilbert space orthogonal complement $\mathbf{M}^{\perp}$, and $L_{k \times n}^2 = \mathbf{M}^{\perp} \oplus \mathbf{M}$ is automatic. Also $\underline{\nu}^- = \{0, \dots, 0\}$. Thus any block invariant subspace of $L_{k \times n}^2$ has a representation as $\underline{\mathbf{M}} = g H_{0\Omega_L}^2(\underline{\nu})$ where $\underline{\nu} = \{\nu_1, \dots, \nu_n\}$ and $\nu_j = \dim \mathbf{M}_j / \mathbf{M}_{j+1}$, and where $g(e^{i\theta})$ is unitary. For $n = 1$, this reduces to the Beurling-Lax theorem; where $\mathbf{M}_1 = H_n^2$ this matches up with the Beurling-Lax theorem for finite matrices formulated in [BG1].

Applications of Theorems 7.4 and 7.5 include analogs of all the applications presented above for the case $n = 1$; the series of papers [BG1–BG4] develops them for the finite matrix setting ($\mathbf{M}_1 = H_k^2$). We also mention that these theorems can be used as tools to understand results from [SW2]; we plan to publish details elsewhere. Here we restrict ourselves to two applications: one is an $H_{n \times n}^\infty$ distance problem; and the other is the positive-definite completion problem discussed in Chapter 9. Finally we present the engineering problem [FeTa] [KPT] which motivates the distance problem.

B. The distance from $L_{n \times n}^\infty$ to $H_{0\Omega_L}^\infty$. In this section we approximate with $H_{n \times n}^\infty$ functions which when evaluated at $z = 0$ have a value which is a lower triangular matrix. This illustrates well the point behind introducing flags; they let us handle the lower triangular at the origin restriction.[1] This mathematical problem first arose in the control context (as in Chapter 1). Feintuch, Khargonekar, Poola, and Tannenbaum pointed out that in the study of periodic systems this problem arose. A solution is given in [FeTa]. We show how this can be done with flags and obtain a parametrization for all solutions.

The problems we shall treat are formally stated as:

PROBLEM 7.A. Given $K \in L_{n \times n}^\infty$, determine if there is a $H \in H_{0\Omega_L}^\infty$ such that $\|K - H\|_\infty \leq 1$.

PROBLEM 7.B. Given $K \in L_{n \times n}^\infty$, parametrize the set $\{F = K - H \colon H \in H_{0\Omega_L}^\infty, \|F\| \leq 1\}$.

To handle these problems, introduce the Hilbert space $H_{0\Omega_L}^2$; the inner product is given by

$$\langle H, G \rangle = \frac{1}{2\pi} \int_0^{2\pi} \operatorname{tr}(G(e^{i\theta})^* H(e^{i\theta})) \, d\theta.$$

Note that $H_{0\Omega_L}^2$ is a module over $H_{0\Omega_L}^\infty$; that is,

$$G \in H_{0\Omega_L}^2, \quad L \in H_{0\Omega_L}^\infty \text{ implies } GL \in H_{0\Omega_L}^2.$$

The module homomorphisms on $H_{0\Omega_L}^2$ (i.e., operators mapping $H_{0\Omega_L}^2$ to itself which commute with the $H_{0\Omega_L}^\infty$-module action of right multiplication) are easily seen to be the operators of *left* multiplication by an $H_{0\Omega_L}^\infty$ function. From this

---

[1] An interpolation enthusiast might wonder why we do not treat such a restriction as an interpolation constraint at 0. The answer is that the restriction is not of the type $(K + \theta H \Psi)$; so standard $\mathbf{M}$, $\mathbf{M}'$ interpolation theory fails.

we see that if $\mathbf{G} = [^F_I]H^2_{0\Omega_L}$ is the graph space of a function $F \in K + H^\infty_{0\Omega_L}$, then

(7.2) $\mathbf{G}$ is $H^\infty_{0\Omega_L}$-invariant

and

(7.3) $$\mathbf{G} \subset \underline{\mathbf{M}} := \begin{bmatrix} K \\ I \end{bmatrix} H^2_{0\Omega_L} + \begin{bmatrix} H^2_{0\Omega_L} \\ (0) \end{bmatrix}.$$

(To be precise we should write $[^{LF}_I]H^2_{0\Omega_L}$ rather than $[^F_I]H^2_{0\Omega_L}$ to emphasize that the action of F on $H^2_{0\Omega_L}$ is that of left multiplication by the matrix function F but we expect no confusion will result; similar remarks apply to $[^K_I]H^2_{0\Omega_L}$.) To analyze the condition $\|F\|_\infty \leq 1$, introduce the indefinite inner product on $L^2_{2n \times n}$ by

$$[G, H]_J = \frac{1}{2\pi} \int_0^{2\pi} \mathrm{tr}(H(e^{i\theta})^* J G(e^{i\theta})) \, d\theta$$

where $J = \begin{bmatrix} I_n & 0 \\ 0 & -I_n \end{bmatrix}$. Then $\|F\|_\infty \leq 1$ means that $\mathbf{G} = \binom{F}{I}H^2_{0\Omega_L}$ satisfies the condition that

(7.4) $\mathbf{G}$ is a maximal J-negative subspace of $\mathbf{K} := \begin{bmatrix} L^2_{n \times n} \\ H^2_{0\Omega_L} \end{bmatrix}$.

Conversely, one can check that if a subspace $\mathbf{G}$ satisfies (7.2), (7.3), and (7.4), then necessarily $\mathbf{G}$ has the form $[^F_I]H^2_{0\Omega_L}$ for an $F \in K + H^2_{0\Omega_L}$ with $\|F\|_\infty \leq 1$.

For there to exist a subspace $\mathbf{G}$ satisfying (7.3) and (7.4) simultaneously, as in Chapter 5, it is necessary and sufficient that $\mathbf{K} \boxminus \mathbf{M}$ be a positive subspace. This is equivalent to the Hankel-type operator $\underline{\mathbf{H}}_K$: $H^2_{0\Omega_L} \to H^{2\perp}_{0\Omega_L}$ given by

$$\underline{\mathbf{H}}_K: \ G \to P_{H^{2\perp}_{0\Omega_L}}(L_K G)$$

having norm ≤ 1. It is not difficult to see that $\underline{\mathbf{H}}_K$ is unitarily equivalent to a direct sum of traditional Hankel operators

$$\underline{\mathbf{H}}_K \cong \begin{bmatrix} \mathbf{H}_{K_1} & & 0 \\ & \ddots & \\ 0 & & \mathbf{H}_{K_n} \end{bmatrix} .$$

where $\mathbf{H}_{K_j}$: $H^2_n \to H^{2\perp}_n$ is the Hankel operator $\mathbf{H}_{K_j}$: $f \to P_{H^{2\perp}_n}(K_j f)$ with symbol

$$K_j = Z_j^* K Z_j \quad \text{where} \quad Z_j = \begin{bmatrix} zI_{j-1} & 0 \\ 0 & I_{n+1-j} \end{bmatrix} .$$

We have thus established: *a necessary condition for a solution to Problem* 7.A *to exist is that* $\|\mathbf{H}_{K_j}\| \leq 1$ *for* $1 \leq j \leq n$. To establish sufficiency, it is convenient to assume that $\|\mathbf{H}_K\| < 1$ (or equivalently, $\|\mathbf{H}_{K_j}\| < 1$ for $1 \leq j \leq n$). The general existence question can then be handled by approximation. Let us also suppose that K is a smooth matrix function on the circle $\{|z| = 1\}$. Then $\mathbf{M}$ is a regular subspace in the J-inner problem and when we apply Theorem 7.5 to $\mathbf{M}$ the representer g turns out also to be smooth, so in particular, $g \in L^\infty_{2n \times 2n}$. Also for our case here, $\underline{\nu}^+ = \{1, \ldots, 1\} = \underline{\nu}^-$. Thus we have the representation for $\mathbf{M}$,

$$\mathbf{M} = g \begin{bmatrix} H^2_{0\Omega_L} \\ H^2_{0\Omega_L} \end{bmatrix},$$

where $g(e^{i\theta})^* J g(e^{i\theta}) = J$. Note that g, as a left multiplication operator, commutes with the $H^\infty_{0\Omega_L}$-module action of right multiplication and g preserves the J-inner product. With the assumption $\|\mathbf{H}_K\| \le 1$, the condition (7.4) can be weakened to

(7.4′) $\mathbf{G}$ is a maximal J-negative subspace of $\mathbf{M}$.

By these considerations we see that a subspace $\mathbf{G}$ satisfies (7.2), (7.3), and (7.4) if and only if $\mathbf{G} = g\mathbf{G}_1$ where

(7.5) $\mathbf{G}_1$ is $H^\infty_{0\Omega_L}$-invariant,

(7.6) $\mathbf{G}_1 \subset \begin{bmatrix} H^2_{0\Omega_L} \\ H^2_{0\Omega_L} \end{bmatrix},$

and

(7.7) $\mathbf{G}_1$ is a maximal J-negative subspace of $\begin{bmatrix} H^2_{0\Omega_L} \\ H^2_{0\Omega_L} \end{bmatrix}.$

But subspaces $\mathbf{G}_1$ satisfying (7.5)–(7.7) are easily analyzed; they are all those subspaces of the form $\mathbf{G}_1 = \begin{bmatrix} G \\ I \end{bmatrix} H^2_{0\Omega_L}$ where $G \in H^\infty_{0\Omega_L}$ and $\|G\|_\infty \le 1$. On the other hand we know $g\mathbf{G}_1 = \begin{bmatrix} F \\ I \end{bmatrix} H^2_{0\Omega_L}$ where F is a solution of Problem 7.A. If we write g out in block matrix form $g = \begin{bmatrix} \alpha & \beta \\ \kappa & \gamma \end{bmatrix}$, we thus have

$$\begin{bmatrix} F \\ I \end{bmatrix} H^2_{0\Omega_L} = \begin{bmatrix} \alpha & \beta \\ \kappa & \gamma \end{bmatrix} \begin{bmatrix} G \\ I \end{bmatrix} H^2_{0\Omega_L}.$$

From this identity it is easy to deduce that $F = (\alpha G + \beta)(\kappa G + \gamma)^{-1}$. This provides a solution of Problem 7.B for the generic case $\|\mathbf{H}_K\| < 1$. We have obtained

THEOREM 7.6. *Problem 7.A has a solution if and only if $\|\mathbf{H}_{K_j}\| \le 1$ for $1 \le j \le n$. If $\|\mathbf{H}_{K_j}\| < 1$ for $1 \le j \le n$, then there is a matrix function $g = \begin{bmatrix} \alpha & \beta \\ \kappa & \gamma \end{bmatrix}$ such that solutions F of Problem 7.A are precisely those functions of the form $F = (\alpha G + \beta)(\kappa G + \gamma)^{-1}$ for a matrix function $G \in H^\infty_{0\Omega_L}$ with $\|G\|_\infty \le 1$.*

REMARKS. For the case $n = 1$, the above construction is done in detail in [**BH2**]. For the case where K is constant, see [**BG1**]; this case gives an alternate proof of the Arveson distance formula for finite nests. The construction is actually more general; one can prove a lifting theorem which contains the Sarason–Sz.-Nagy–Foiaş lifting theorem and the lifting theorem for finite matrices from [**BG2**] as special cases.

C. Positive definite completions. The following problem was first studied by H. Dym and I. Gohberg.

PROBLEM 7.C. Given a band matrix $P = [P_{ij}]_{1 \leq i,j \leq n}$ (where $P_{ij} = 0$ if $|i - j| \geq m$), determine if there is a matrix $Q = [Q_{ij}]_{1 \leq i,j \leq n}$ with $Q_{ij} = 0$ if $|i - j| < m$ such that $P + Q$ is positive definite.

This is the band case of the more general problem solved by Grone, Johnson, de Sá, and Wolkowicz (see Chapter 9). We indicate here how this case can be handled by invariant subspace representation techniques. As a bonus we get a linear fractional representation for the set of all solutions; this is all discussed in [**BG1**] and [**BG2**]. An open question is to see if the general case in [**GJSW**] can be handled by some version of these techniques.

We first reformulate the problem. Write $P = K + K^*$ for a matrix $K \in \Omega_L$. Let Ω_L^m be the subspace of Ω_L consisting of matrices $[F_{ij}]_{1 \leq i,j \leq n}$ such that $F_{ij} = 0$ for $i > j + m$. The Problem 7.C can be reformulated as

PROBLEM 7.D. Given $K \in \Omega_L$, determine if there is a matrix $L \in \Omega_L^m$ such that $(K + L) + (K + L)^*$ is positive definite.

The basic observation is that a matrix F in Ω_L has a positive definite real part if and only if its graph $\mathbf{G} = [\begin{smallmatrix} F \\ I \end{smallmatrix}]\Omega_K^2$ is a strictly J'-negative maximal J'-negative subspace of $[\begin{smallmatrix} \Omega_L^2 \\ \Omega_L^2 \end{smallmatrix}]$, where $J' = [\begin{smallmatrix} 0 & -I_n \\ -I_n & 0 \end{smallmatrix}]$. Here Ω_L^2 is the set Ω_L of lower triangular matrices, considered as a Hilbert space in its Hilbert-Schmidt norm. To handle Problem 7.D one need only go through the analysis above for Problems 7.A and 7.B, but with J' in place of J and with all functions restricted to be constants. It turns out that the obvious necessary condition for solutions to exist, that all square submatrices of $P = K + K^*$ in the band area $\{(i, j): |i - j| < m\}$ be positive definite, is also sufficient. Moreover, Theorem 7.4 applied to this setting can be used to get a linear fractional parametrization of the set of all solutions. The result can also be seen as an application of the lifting theorem for finite matrices; this is explained in [**BG2**].

REMARKS. Note that actually we have the machinery to handle a more general problem corresponding to letting K be a function in $H^2_{0\Omega_L}$ and replacing Ω_u^m by some $H^\infty_{0\Omega_L}$-invariant subspace of $H^2_{0\Omega_L}$ (e.g., $z^m H^2_{0\Omega_L}$). In terms of the function $P = K + K^*$, we specify the Fourier coefficients P_j of P for $|j| < m$, specify only the lower triangular part of the coefficient P_m, and hence also only the upper triangular part of $P_{-m} = P_m^*$; the question then is: Can the upper triangular part of P_m (and hence also the lower triangular part of P_{-m}) together with coefficients P_j for $|j| > m$ be chosen so that the resulting function $P(z)$ has positive definite values on the unit circle? Again the natural necessary conditions turn out also to be sufficient. This problem has been studied at length by Dym and Gohberg for the case where full coefficients P_j are prescribed for $|j| \leq m$.

D. A control problem. Consider the following discrete time linear feedback system.

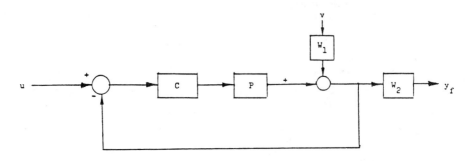

Here we assume that P is an n-periodic plant, and W_1 and W_2 are stable n-periodic operators with stable inverses. For simplicity we assume that the system and all its components are single-input single-output. Then all operators C, P, W_1, W_2 we consider as being operators on L^2. The operator X is said to be periodic if X commutes with the nth power M_{z^n} of the shift operator M_z of multiplication by z. We assume that P is strictly causal; that is, $PH^2 \subset zH^2$. The operator X is stable if $XH^2 \subset H^2$. The objective is to design C to minimize the energy of y_f for the worst v of unit energy, where C is constrained to be causal, n-periodic, and to achieve internal stability. Thus the cost is

$$\text{cost} = \sup\{\|y_f\|_2 : v \in H^2, \|v\|_2 \le 1\}.$$

In terms of the transfer matrix v to y_f, we have $\text{cost} = \|W_2(I + PC)^{-1}W_1\|$.

To cast the problem into more familiar terms, we map H^2 into H_n^2 by the following rule. An f in H^2 has Fourier coefficients $(f_1, f_2, f_2, \ldots)$. The corresponding F in H_n^2 is defined to have Fourier coefficients

$$\begin{pmatrix} f_1 \\ f_2 \\ \vdots \\ f_n \end{pmatrix}, \begin{pmatrix} f_{n+1} \\ \vdots \\ f_{2n} \end{pmatrix}, \begin{pmatrix} f_{2n+1} \\ \vdots \\ f_{3n} \end{pmatrix}, \ldots.$$

Under this mapping n-periodic operators on L^2 correspond to operators on L_n^2 which commute with the M_z on L_n^2, that is, to left multiplication by an $L_{n \times n}^\infty$ function. The causal stable operators correspond to multiplication by an $H_{n \times n}^\infty$ function F such that $F(0)$ is lower triangular, i.e., to left multiplication by an $H_{0\Omega_L}^\infty$ function. After some further manipulation (using the Youla parametrization of all stabilizing compensators and some matrix inner-outer and coprime factorizations), it is shown in [**FeTa**] that the sensitivity minimization problem given above reduces to a problem of the form

$$\text{minimize } \{\|K - H\|_\infty : H \in H_{0\Omega_L}^\infty\}$$

where $K \in L_{n \times n}^\infty$ is a known function. This is indeed Problems 7.A and 7.B.

Appendix to Part II:
Remarks on Interpolation and Operator Theory

Interpolation theory began with Carathéodory and Fejer around 80 years ago. Serious contributions were made by many analysts including Pick, Nevanlinna, Löwner, Takagi, and Nehari. In the 1950s attention turned to interpolation with matrix-valued functions and this subject was the province of operator theorists. Nagy and Koranyi [**NK**] gave the first approach in the middle fifties, and Sarason [**S**] in 1967 connected interpolation with the problem of lifting two commuting contractions (commutant lifting) and proved lifting in a special case. Nagy and Foiaş (cf. [**NF**]) proved lifting in general and thereby established an extremely general interpolation theorem. It was later realized that the NF lifting theorem followed from an earlier theorem of Ando. Over the next ten years it was established that commutant lifting had many consequences, principally for interpolation, H^∞ approximation, and the Toeplitz Corona Theorem. That is where the idea originates, which is emphasized in these notes, that many theorems follow from one general principle.

Almost simultaneously Adamjan, Arov, and Krein [**AAK**] performed an extensive study of matrix interpolation and approximation from $H^{\infty;l}$. D. Clark [**Cl**] independently obtained overlapping results. Many extensions and refinements were obtained by mathematicians Arsene-Ceaşescu-Foiaş [**ACF**], Ball-Helton [**BH**], Dym-Gohberg [**DG**], Fedchin [**Fd**], Ivanchenko [**Iv**], Nudelman [**Nu**] and Kovaluhyn-Potopov [**KP**] of the Ukrainian school, and by engineers DelSarte-Genin-Kamp [**DGK**], DeWilde and Dym [**DD**], Doyle [**Doy**], Francis-Helton-Zames [**FHZ**], Glover [**Gl**], Jonckheere [**J**], S. Y. Kung [**K**], and Betayeb-Silverman [**BSS**]. There are earlier and parallel developments in closely related areas. One is extensive work (during the 1960s) of Potopov [**EP, P**] on J-unitary valued matrix functions which picks up many of the interpolation results mentioned here. We mention also recent work [**A-D**] of Alpay and Dym on reproducing kernel Krein spaces which is related to the **M**, **M′** representation theorem (Theorem 4.1). This extends work on reproducing kernel Hilbert spaces by de Brangés in the early 1960s [**deBr1**], [**deBr2**]. See also recent works [**deBr4–deBr5**].

Chapter 7 describes a blending of the function theory already described with operator theoretic work on chains and nests (flags). Operator theoretic work even in the 1970s permitted nests of infinite length. Gohberg and Barker [**GB**] studied block LU factorization for infinite matrices. In 1975 Arveson computed the distance of a block matrix to block upper triangular matrix, and gave a block Toeplitz Corona Theorem [**A**]. Subsequently Feintuch [**Fe**] developed matrix interpolation.

Ball-Gohberg [**BG**] converted the invariant subspace approach of [**BH**] for analytic functions to upper triangular matrices. If the results of Chapter 7 are specialized to constant functions, they give exactly the Ball-Gohberg results. Conversely, (omitted) proofs of the main theorems in Chapter 7 can be gotten by combining arguments in [**BG**] and [**BH**]. Details are planned for a future article by Ball-Gohberg-Helton.

Another component of Chapter 7 is the work of Segal and Wilson. Lie group investigators have used flags heavily for so many years that I would not venture to say who in that community did the first serious work with them. From the viewpoint of "Wiener-Hopf factorization" Segal and Wilson [**SW**] show the existence of a factor A_+ which is lower triangular at the origin. This was done with a different proof by Goodman and Wallach [**GW**]. Wiener-Hopf factorization had previously been connected with integrable systems, and the key insight in [**SW**] is that an invariant subspace way of handling "Wiener-Hopf factorization" elucidates many of the computations with integrable systems.

Historically the many items presented in one framework in Chapters 5, 6, and 7 can be thought of in clusters. First is the commutant lifting cluster, second the "Wiener-Hopf" factorization, while third is the "Wiener-Hopf" treatment of integrable systems. Possibly the main new contribution of the Ball-Helton approach is that it ties commutant lifting cluster intimately to Wiener-Hopf factorization [**BH1,2,8**]; consequently all items in all three areas are united. Also the thoroughly geometric nature of our proofs is new for commutant lifting related theorems and for some types of "factorization."

Besides general remarks we give some detailed attributions. Chapter 4 is a proper subset of [**BH2**]. Theorem 5.1 is from [**BH1**]. The results of A in Chapter 5 on interpolation are due to Pick, Nevanlinna, and Takagi; the proofs given are from [**BH1**]. In B of Chapter 5 in Corona theory, Theorem 5.6 (for $l = 0$, but without the $\mathbf{G}_g$ parametrization) was first given in [**NF2**]. The parametrization was suggested by Foiaş in his 1979 Toeplitz lecture. The proof given here is immediate from the Ball-Helton machine and first appears (along with the $l \neq 0$ results) in [**H10**]. Chapter 5.C on factorization already contains references. An added point is that factorization Theorem 5.8 for $A \geq 0$ is one of the traditional successes of the classical Beurling-Lax theorem (see [**NF1**] or [**RR1**]). The results on "disk problems" are just a simple variant on standard results on H^∞-approximation. Two physicists around 1970 used these results to fit scattering data. Helton [**H4**] and [**H5**], unaware of this earlier application, proposed using disk results in amplifier design.

Section 6.A is due to Ball-Helton and is new, Section 6.B is from [**BH3**], Section 6.C is from [**BH1**], Section 6.D is from [**BH1,5**], Section 6.E is from Chapter 6 [**D**]. Section 6.F contains references. Chapter 7 was written jointly by Ball and Helton and is new.

Several *surveys in closely related areas* are:

J. Ball has written a survey on interpolation [**B3**].

H. Dym, NSF Regional Conference (1984) on J-unitary matrix functions.

B. Francis and J. Doyle [**FD**] are writing a survey on control theory aspects of this mathematics.

B. Francis, the best book on control aspects of this subject [**Fr**].

R. Herman is writing a book on the subject.

M. Rosenblum and J. Rovnyak have a book on function theory derivable with operator techniques [**RR2**].

D. Sarason, Lecture Notes from 1984 Lancaster conference [**S**].

G. Segal and G. Wilson are writing a book on their approach to KdV.

N. Young, Lecture Notes from 1984 Lancaster conference [**Y**].

Software to compute H^∞ interpolants and best $H^{\infty;l}$ approximation has been developed by several groups. The programs of Chang–Pearson, Helton–Schwartz, Kwakernaak, Limebeer, and Postlethwaite are portable and one can get a tape by asking. The computational efforts I know involving $H^{\infty;l}_{m \times n}$ are:

Alison-Yeh-Young (Univ. of Glasgow; Math.)	Pascal $m = n = 1$, $l = 0$. Algol matrix case $l = 0$.
Chang-Pearson (Rice; EE)	Fortran, $l = 0$.
Chu-Doyle (Honeywell, Cal Tech; EE)	Fortran state space Also complicated Γ's as in Ch. 11.
Foiaş (Indiana Univ.; Math)	IBM PC, $l = 0$; uses his "choice sequences."
Francis (Toronto; EE)	$m = n = 1$, $l = 0$.
Glover (Cambridge; EE)	State space.
Trefethn (Yale; CS)	Fortran $m = n = 1$.
Helton-Schwartz (UCSD; Math.)	Fortran $n = 1$, $l = 0$. Also complicated Γ's as in Ch. 11.

Kwakernaak (Twente; Math)	$m = n = 1,\ l = 0$
Limebeer (Imperial College; EE)	State space.
Postlethwaite (Oxford; EE)	State space.
F. B. Yeh (Taiwan Univ. Aeronautics and Astronautics Institute)	Fortran $n = 1,\ l = 0$; also $m \times n$ by S. Y. Kung method.

Different programs input data in different ways. The older efforts were inspired by circuit theory rather than control.

Part III. Matrices

This part deals with approximation problems in spaces of matrices and spaces of operators. Physically these arise in many ways and we shall ultimately list several. Two problems of the type we confront in Part III are:

Define a *disk* Δ_Y in matrix space by $\Delta_Y = \{w \in M_{mm}: w^*aw + w^*f + fw + d$ is a negative semidefinite matrix$\}$ where $Y = \left[\begin{smallmatrix} a & f \\ f_* & d \end{smallmatrix}\right]$ is selfadjoint. Given a disk Δ_Y of $m \times n$ matrices and a real subspace $\mathbf{S}$ of the $m \times n$ matrices:

(P1)   *Do the two intersect? Find a matrix in the intersection, $\mathbf{S} \cap \Delta_Y$, or compute all such matrices.*

(P2)   *Given a set $\mathbf{P}$ of admissible perturbations and a matrix w inside $\mathbf{S} \cap \Delta_Y$: Is $(w + \mathbf{P}) \cap \mathbf{S}$ fully contained in Δ_Y?*

The disk Δ_Y, the subspace $\mathbf{S}$, and $\mathbf{P}$ are given algebraically. How does one actually compute an answer to (P1) and (P2)? Of course any complex subspace $\mathbf{S}$ is also a real subspace.

In order to hammer in the nature of problem (P1) we give the reader a brief test. Brief it is: determine within 15 seconds if the line

$$2x + 3y = 5$$

intersects the disk

$$2x^2 + x + y^2 \le 3.$$

Tick, tick, tick; time is up. The point is that all information is given algebraically, and when we do this in many variables, serious computational problems emerge.

Both problems are exceedingly general; so much so that we shall be delighted whenever we can find a subspace $\mathbf{S}$ or a class of perturbations $\mathbf{P}$ for which any analytic results are possible.

Chapters 8 and 9.C address problem (P1), while Chapter 10 discusses problem (P2).

8. Some Matrix Problems in Engineering

This chapter is motivational and sketches how problem (P1) can arise in engineering design. The problem (P1) occurs in optimal parameter selection, the problem (P2) checking that a design is robust to uncertainties. This chapter is essentially a prelude to the next chapters in Part III and no substantial theorems are proved.

First some mathematical perspective. Consider a more constrained version of (P1). Analyze the intersection

$$(\text{P}'1) \qquad \eta \triangleq \mathbf{S} \cap \Delta_{Y^1} \cap \Delta_{Y^2} \cap \cdots \cap \Delta_{Y^L}$$

of sets in $M_{m \times n}$. We now demonstrate that (P'1) on $M_{m \times n}$ is equivalent to (P1) on a bigger space.

This comes from a simple direct sum construction. For w in $M_{m \times n}$ define

$$(8.1) \qquad D_w = \left. \begin{pmatrix} w & 0 & 0 & 0 \\ 0 & w & 0 & 0 \\ 0 & 0 & \ddots & \\ 0 & 0 & & w \end{pmatrix} \right\} L$$

and

$$Y = \begin{pmatrix} Y_{11}^1 & & & Y_{12}^1 & & \\ & Y_{11}^2 & & & Y_{12}^2 & \\ & & \ddots & & & \ddots \\ Y_{21}^1 & & & Y_{22}^1 & & \\ & Y_{21}^2 & & & Y_{22}^2 & \\ & & \ddots & & & \ddots \\ & & & & & Y_{22}^L \end{pmatrix}.$$

Set $\tilde{\mathbf{S}} = \{D_w \colon w \in \mathbf{S}\}$. Then $D_w \in \tilde{\mathbf{S}} \cap \Delta_Y$ says

$$w^* Y_{11}^j w + w^* Y_{12}^j + Y_{21}^{j*} w + Y_{22}^j \leq 0 \quad \text{for } j = 1, \ldots, L,$$

which says

PROPOSITION 8.1. $m \in \eta$ if and only if $D_w \in \tilde{\mathbf{S}} \cap \Delta_Y$.

This proposition implies that (P1) is an extremely general problem in the following sense. Suppose that we have constraints which give compact convex sets $C^1, \ldots, C^T$ in $\mathbb{C}^n$ and $\mathbf{S}$ is a subspace of $\mathbb{C}^m$. The problem of analyzing

(8.2) $$\mu \triangleq \mathbf{S} \cap C^1 \cap C^2 \cap \cdots \cap C^T$$

is within ε of a (P1) problem on a large space.

PROPOSITION 8.2. Identify $\mathbb{C}^m$ with the $m \times 1$ matrices. Given $\varepsilon > 0$, there is a matrix space $M_{Lm \times L}$, a subspace $\tilde{\mathbf{S}}$ of it, and a convex disk Δ_Y in it, so that if $w \in \mu$, then D_w is within distance ε of $\tilde{\mathbf{S}} \in \Delta_Y$.

Here $D_w =$ the L-fold block diagonal matrix as in (8.1).

PROOF OF PROPOSITION 8.1. Since μ is convex, it is the intersection of half spaces. The intersection of a finite number of these half spaces approximates it to within ε. A very large disk Δ_Y approximates a half plane. Thus μ is within ε of the intersection of a finite number of disks, and so there is an integer $L > 0$ for which a set of the form η approximates μ to within ε. Apply the direct sum construction in Proposition 8.1 to η.   Q.E.D.

Recall that (exterior) nonconvex disks Δ_Y can occur. In fact any reasonable set η can be approximated using a Δ_Y (which may be nonconvex). This and the proposition are compelling evidence that we shall ultimately need more restrictions on $\mathbf{S}$ in (P1) to get strong results. In fact one might ask: Why state (P1) in terms of matrix balls rather than as convex sets? The reason is that our physical problems are presented in the form (P1) or (P1'). Secondly, while a theoretician can easily move to a much bigger space, numerically this can be a disaster; so, the equivalence given in Proposition 8.1 may not be all that useful.

Now we mention some engineering motivation. One way in which (P1) occurs in engineering goes much like the prototypical example in Chapter 2. In designing a system, as in Figure 2.1, one wants to optimize performance $\Gamma(\omega, z(i\omega))$ over designable parameters z. Before we emphasized problems where one designed for good performance over all frequencies. This is called *broadband* design. We emphasized taking z's to be matrix-valued functions, for example in $H^\infty_{m \times n}$.

Narrowband design refers to achieving performance at only one frequency ω_0. As z sweeps through $H^\infty_{m \times n}$ the values $z(i\omega_0)$ sweep through the $m \times n$ matrices. Thus a typical optimization problem is to find

(8.3) $$\{z \in M_{m \times n} : \Gamma(\omega_0, z) \le c\};$$

is it empty? The Γ's which arise have a second form which comes from a universal engineering principle.

You can express Figure 2.1 as shown in Figure 8.1. That is, the designable parameters z can be "extracted" from the system. By Chapter 3 the transfer

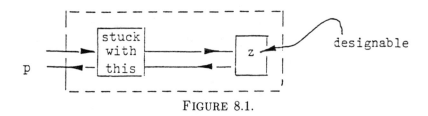

FIGURE 8.1.

function p for the entire system can be written $p = \mathbf{F}_{\left(\begin{smallmatrix} ab \\ cd \end{smallmatrix}\right)}(z)$ where a, b, c, d are determined by the part you are stuck with. If performance is measured by

$$|p|^2, \quad |p - e|^2, \quad |p - e|^2 + |p - r|^2, \text{ etc.,}$$

then Γ is

$$|\mathbf{F}_{\left(\begin{smallmatrix} ab \\ cd \end{smallmatrix}\right)}(z)|^2, \quad |\mathbf{F}_{\left(\begin{smallmatrix} ab \\ cd \end{smallmatrix}\right)}(z) - e|^2, \text{ etc.,}$$

and its sublevel sets $\mathbf{S}_\omega(c)$ being sublevel sets of L.F.T.'s, are disks or limits of them in a fairly weak sense.[1] Since different performance measures $\Gamma_1, \Gamma_2, \ldots, \Gamma_L$ give different disks, the question of deciding those z for which $\Gamma_j(\omega, z) \le c_j$ for all j is the same as the problem of finding those matrices z which are in the intersection

(8.4) $$\Delta_{Y^1} \cap \Delta_{Y^2} \cap \cdots \cap \Delta_{Y^L}.$$

This, of course, is a special case of (P'1). *The z's which give performance within a specified range (meet the specs) are exactly a set of the form* (8.4).

The rather basic design problem we have been discussing leads us directly to problem (P'1) where the set $\mathbf{S}$ is selected to be $M_{m \times n}$. How do other sets arise?

(i) Sometimes one wants various of the inputs and outputs of the designable z not to interact with each other. Mathematically, this means that we insist that all admissible z have certain entries equal to zero. So $\mathbf{S}$ consists of all matrices which have a given sparsity pattern. For example, $\mathbf{S} =$ the diagonal $n \times n$ matrices is common; this occurs when each input is paired with each output, and one refers to this as a decoupled situation.

(ii) Another situation involves $z \in \mathbb{R}^n \subset \mathbb{C}^n$ which one could identify with the $n \times 1$ matrices. The whole idea is explained by one example. Suppose we are building an electronic circuit whose basic layout (called "topology" by circuit theorists) has been decided, only the capacitor, inductor, and resistor values are yet to be selected. Since these are real we get a real subspace, so $\mathbf{S} = \mathbb{R}^n$.

---

[1]For matrix a, b, c, d, z this statement about disks holds in the following sense:

LEMMA 8.3. *Any sublevel set* $\mathbf{S}(c) = \{z \colon \|\mathbf{F}_{\left(\begin{smallmatrix} ab \\ cd \end{smallmatrix}\right)}(z)\| \le c\}$ *of* $\mathbf{F}$ *is a matrix disk* Δ_Y *provided that either b or c is invertible. Even if neither b nor c is invertible, there is a family of disks* Δ_{Y_ε} *with Y_ε given by formula* (8.11) *so that the set* $\mathbf{S}(c)$ *is approximated by* Δ_{Y_ε} *when ε is near 0.*

PROOF. See Appendix to Chapter 8.

(iii) A very different way (P1) occurs is in connection with time varying circuits. These causal systems correspond to upper triangular infinite matrices. Control problems like those in Chapter 2 ultimately require approximation by H^∞ functions, $\inf_{f \in H^\infty} \|g - f\|_{L^\infty}$, which is the same as

$$\inf_{f \in H^\infty} \|T_g - T_f\|_{\mathbf{L}(H)},$$

where T_h is the Toeplitz operator generated by h. Now $f \in H^\infty$ is equivalent to T_f upper triangular. The Chapter 2 type of control for time varying systems ultimately requires approximation by infinite upper triangular matrices

$$(8.5) \qquad\qquad \inf_{F \text{ upper } \Delta} \|G - F\|_{\mathbf{L}(H)}.$$

The last problem lies in the realm of success for the theory we are about to present. Indeed the theory gives a fairly complete understanding of problem (8.5). We treat the earlier problems less successfully here.

Least successful is our approach to the problem of a real subspace. It does not fall into our purvey at all and we will not mention it again. There is, of course, extensive literature on circuit optimization and we don't attempt to review it.

Chapters 7 and 9 address problem (P1) when $\mathbf{S}$ is the set of all matrices which have a prescribed sparsity pattern.[2] We find explicit formulas for solution of (P1) for quite a few patterns of sparsity. This is a substantial accomplishment, since the results are hard won and many patterns of sparsity defy analysis.

Chapters 7 and 9 represent two completely different approaches. It would be valuable to unify them. The Chapter 7 flag approach in finite dimensions (no $e^{i\theta}$) is due to Joe Ball and Israel Gohberg [**BG1-BG4**]. They make ingenious use of invariant subspaces in the spirit of Part II to derive matrix (or operator algebra) versions of the main function theory results in Part II. This was one of the basic special cases subsumed in Chapter 7.

The second approach (Chapter 9.C) is due to Robert Grone, Charles Johnson, Eduardo de Sá, and Henry Wolkowicz [**GJSW**]. It addresses the following problem:

(8.6)   *Given a matrix $w \in M_{n \times n}$ set $\mathbf{S}$ = all matrices with a prescribed sparsity pattern. Does some $F \in \mathbf{S}$ make $\mathrm{Re}(W - F)$ negative definite?*

The [**GJSW**] approach is definitive. It gives necessary and sufficient conditions on a sparsity class $\mathbf{S}$ for (8.6) to be explicitly solvable by Gauss elimination.

The analog is

(8.7)   is $\|W - F\| \leq 1$ for some F in $\mathbf{S}$?

---

[2]Star pattern to some operator theorists.

The Arveson distance formula generalized as in Chapter 7 by Ball and Gohberg handles $\mathbf{S}$ which are upper triangular matrices in both problems (8.6) and (8.7). Exactly which sparsity classes $\mathbf{S}$ give a "natural" solution to problem (8.7) has recently been discovered [K-L, J-R], and is described in Theorem 9.3. This was inserted subsequent to 1985 when these lectures were originally given.

Appendix: Proof of Lemma 8.3. We have $a \in M_{lk}$, $b \in M_{lt}$, $c \in M_{rk}$, $d \in M_{rt}$, and $z \in M_{tr}$. Suppose c is invertible. We must characterize those z satisfying

(8.8)
$$\mathbf{F}_{\left(\begin{smallmatrix} a & b \\ c & d \end{smallmatrix}\right)}(z)^* \mathbf{F}_{\left(\begin{smallmatrix} a & b \\ c & d \end{smallmatrix}\right)}(z) \le I,$$
$$[a^* + c^*(I + dz)^{-1*}z^*b^*][a + bz(I + dz)^{-1}c] \le I,$$
$$a^*a + c^*(I + dz)^{-1*}z^*b^*a + a^*bz(I + dz)^{-1}c$$
$$+ c^*(I + dz)^{-1*}z^*b^*bz(I + dz)^{-1}c \le I.$$

This equals

$$(I + dz)^*k^*a^*ak(I + dz) + z^*b^*ak(I + dz) + (I + dz)^*k^*a^*bz + z^*b^*bz$$
$$\le (I + dz)^*k^*k(I + dz)$$

with $k = c^{-1}$. More algebra gives

(8.9)
$$k^*a^*ak - k^*k + z^*\{d^*k^*a^*ak + b^*ak - d^*k^*k\}$$
$$+ \{k^*a^*akd + k^*a^*b - k^*kd\}z$$
$$+ z^*\{d^*k^*a^*akd + b^*akd + d^*k^*a^*b + b^*b - d^*k^*kd\}z \le 0.$$

Thus (8.8) is equivalent to z lying in a disk given by (8.9).

The computation when b is invertible is based on analyzing $F(z)F(z)^* \le I$ and is very similar.

When neither b nor c is invertible one can select one of them and make a small modification. Suppose for simplicity that null c is $\{0\}$. Let η be an isometry of $\mathbf{C}^{r-k}$ to (range c)$^\perp$; then $c_\varepsilon \triangleq c + \varepsilon\eta: \mathbf{C}^r \to \mathbf{C}^r$. Let k_ε denote the inverse of c_{ε^2}.

If z satisfies (8.8), then for all ε small enough

(8.10)
$$\mathbf{F}_{\left(\begin{smallmatrix} a & b \\ c_{\varepsilon^2} & d \end{smallmatrix}\right)}(z)^* \mathbf{F}_{\left(\begin{smallmatrix} a & b \\ c_{\varepsilon^2} & d \end{smallmatrix}\right)}(z) \le (1 + \varepsilon)I$$

since

$$3\varepsilon^2\|bz(1 + dz)^{-1}\eta\| \le \varepsilon$$

for small ε. Conversely, if (8.10) holds for a particular z and all small ε, then take $\lim_{\varepsilon \to 0}$ to see that (8.8) holds. Since c_ε is invertible the computation leading to

(8.9) is valid and gives that $S(c)$ is a "limit" of disks Δ_{Y_ε} where Y_ε is given by

(8.11)
$$
Y_\varepsilon = \begin{pmatrix} k_\varepsilon(a^*a - [1+\varepsilon])k_\varepsilon & \{d^*k_\varepsilon^*a^*a - d^*k_\varepsilon^*[1+\varepsilon] + b^*a\}k_\varepsilon \\ k_\varepsilon^*\{a^*ak_\varepsilon d - [1+\varepsilon]k_\varepsilon d + a^*b\} & \begin{matrix} d^*k_\varepsilon^*a^*ak_\varepsilon d + b^*ak_\varepsilon d + d^*k_\varepsilon^*a^*b \\ + b^*b - d^*k_\varepsilon^*[1+\varepsilon]k_\varepsilon d \end{matrix} \end{pmatrix}.
$$

We remark that while approximation of Δ_Y with Δ_{Y_ε} appears to be in a strong sense, some unusual phenomena are readily possible. Namely, when a, b, c are rank one and z is restricted to lie in the one-dimensional space of matrices $\mathbf{S} = \{zI\}$, then $\mathbf{F}_{\left(\begin{smallmatrix} a & b \\ c & d \end{smallmatrix}\right)}(zI)$ can be selected to be any rational function (by a fundamental theorem of system theory); consequently, $\mathbf{S} \cap \Delta_Y$ can be extremely pathological.

9. Optimization, Matrix Inequalities, and Matrix Completions

The ideas of optimization have become a significant tool in bounding matrix parameters and conveying insight into matrix problems generally. We outline here a general approach to bounding matrix parameters (§B) and discuss some recent results on matrix completion problems in which optimization plays a natural role (§C). We begin with a brief discussion of some classical inequalities as background (§A).

A. The Hadamard/Fischer inequalities and generalizations. Our purpose here is (i) simply to note for reference some classical determinantal inequalities for positive definite matrices, and (ii) to observe that they are ultimately consequences of a familiar inequality for positive numbers. Though sometimes subtle, this connection of matrix parameter inequalities to inequalities/optimization for numbers is common.

Let $A = (a_{ij})$ be an n-by-n positive definite Hermitian matrix throughout this section, and let $\alpha, \beta \subseteq \{1, 2, \ldots, n\}$ be index sets. By $A[\alpha]$ we mean the principal submatrix of A contained in the rows and columns indicated by α, while $A(\alpha)$ denotes the complementary principal submatrix of A resulting from deletion of the rows and columns indicated by α; i.e., $A[\alpha'] = A(\alpha)$.

The classical inequalities we wish to note are the following:

$$(9.1) \qquad \det A \le \prod_{i=1}^{n} a_{ii} \qquad (\textit{Hadamard's Inequality});$$

$$(9.2) \qquad \det A \le \det A[\alpha] \det A(\alpha) \qquad (\textit{Fischer's Inequality});$$

$$(9.3) \qquad \det A[\alpha \cup \beta] \le \frac{\det A[\alpha] \det A[\beta]}{\det A[\alpha \cap \beta]} \qquad (\textit{Hadamard/Fisher});$$

and

$$(9.4) \qquad \left(\prod_{|\alpha| = k+1} \det A[\alpha] \right)^{1/\binom{n-1}{k}} \le \left(\prod_{|\alpha| = k} \det A[\alpha] \right)^{1/\binom{n-1}{k-1}},$$
$$k = 1, \ldots, n-1 \qquad (\textit{Szasz's Inequality}).$$

It is a standard and straightforward exercise to see that (9.3) implies (9.2), (9.2) implies (9.1), and (9.4) implies (9.1). (Note that $\det A[\varnothing]$ is taken to be 1 by convention.) The Hadamard inequality (9.1) is equivalent to the statement that, for a general square matrix, the absolute value of the determinant is bounded by the product of row lengths, which bears the same name. Equality occurs in (9.1), for positive definite (i.e., nonsingular) A if and only if A is diagonal.

One of many proofs [J] of (9.1) shows that it is a consequence of the arithmetic/geometric mean inequality for positive numbers. Note that the replacement of A by DAD for $D = \mathrm{diag}(a_{11}^{-1/2}, a_{22}^{-1/2}, \ldots, a_{nn}^{-1/2})$ multiplies both sides of (9.1) by the same positive factor and retains positive definiteness. Thus to prove (9.1) we may assume that each diagonal entry of A is 1. Then, the arithmetic mean of the eigenvalues of A is 1, while the geometric mean of the eigenvalues is $(\det A)^{1/n}$. Thus, since the eigenvalues of A are positive, the arithmetic/geometric mean inequality applied to them yields $(\det A)^{1/n} \leq 1$ or $\det A \leq 1$ which is (9.1).

Actually, each of the generalizations (9.2), (9.3), and (9.4) may be deduced from (9.1), so that they too are consequences of an inequality for numbers. To see that (9.2) follows from (9.1), let U_1 be a $|\alpha|$-by-$|\alpha|$ unitary matrix such that $U_1^* A[\alpha] U_1$ is diagonal and let U_2 be a $(n - |\alpha|)$-by-$(n - |\alpha|)$ unitary matrix such that $U_2^* A(\alpha) U_2$ is diagonal. Then define an n-by-n unitary matrix U with $U[\alpha] = U_1$ and $U(\alpha) = U_2$ (and 0's in remaining entries). It follows that $(U^* A U)[\alpha]$ is diagonal with $\det(U^* A U)[\alpha] = \det U_1^* A[\alpha] U_1 = \det A[\alpha]$ and similarly for $A(\alpha)$, and we have

$$\det A = \det U^* A U \leq \det(U^* A U)[\alpha] \det(U^* A U)(\alpha) = \det A[\alpha] \det A(\alpha).$$

The inequality is a consequence of (9.1) since both submatrices are diagonal.

The inequality (9.3) follows in turn from (9.2) via the special case of Jacobi's determinantal identity for principal minors:

$$(9.5) \qquad\qquad \det A^{-1}(\alpha) = \frac{\det A[\alpha]}{\det A}$$

(see [HJ]). We may suppose without loss of generality that $\alpha \cup \beta = \{1, 2, \ldots, n\}$ and use the fact that the positive definite matrices are closed under inversion. Taking A to be the positive definite matrix $A^{-1}[\alpha' \cup \beta']$ and applying (9.2) yields (because $\alpha' \cap \beta' = \varnothing$) $\det A^{-1}[\alpha' \cup \beta'] \leq \det A^{-1}[\alpha'] \det A^{-1}[\beta']$, and application of the identity (9.5) to each term produces

$$\frac{\det A[\alpha \cap \beta]}{\det A} \leq \frac{\det A[\alpha]}{\det A} \frac{\det A[\beta]}{\det A}.$$

But, multiplication of both sides by $\det A$ then implies (9.3) because $\det A = \det A[\alpha \cup \beta]$.

A proof of (9.4) from (9.1) is given in [HJ, p. 479], again by applying (9.1) to principal submatrices of A^{-1}.

More intricate determinantal inequalities and special eigenvalue inequalities for Hermitian matrices have been deduced entirely from inequalities for real numbers obtained from Lagrange multipliers, for example in [GJSW2] and [WS1-WS2].

B. A general approach to bounding matrix parameters via optimization. We outline here a natural and surprisingly useful technique for obtaining (usually best possible) bounds for matrix parameters via optimization. Suppose we wish to bound some scalar-valued parameter which is a function of a matrix (such as the determinant, permanent, spectral radius, smallest eigenvalue, spread, largest singular value, or numerical radius, etc.) and we wish to use only certain (usually simple) information about the matrix (such as absolute row sums or lengths or other summary information about rows and/or columns). For most parameters and most types of information, the best possible bound is then a solution to a well-posed optimization problem.

Suppose $p(A)$ is the value of the parameter of interest at the matrix of interest A. Let Γ denote the set of all matrices which agree with A with respect to the information to be used (e.g., all matrices with the same absolute row sums as A if the estimate of $p(A)$ is to be based upon absolute row sums). If p depends continuously upon A (a typical situation) and Γ is a compact set (also typical), then the maximum (or minimum) value $p_{\max}$ (or $p_{\min}$) of p over Γ is attained by some matrix in Γ. This value is then the best possible upper (or lower) bound for $p(A)$ if we are to use only about A the information which determines Γ. It is often, but by no means always, the case that important structure of *some* optimizing matrices may be inferred from analysis of the existence of an optimum. Often this structure is combinatorial in nature and permits a finite subset of Γ (determined by "nice" structure) to be analyzed for the optimum. The analysis of this finite set may range from relatively easy to still very difficult. It is worth noting that this finite set usually consists of extreme points of Γ, in spite of a lack of convexity or concavity of p. (Actually, there is often a generalization of convexity involved which still implies an optimum at an extreme point, but we explore this elsewhere.)

To illustrate the approach we have outlined, we mention a couple of basic examples, neither of which is new (except, perhaps, from this point of view).

EXAMPLE 1. Let $A = (a_{ij})$ be an n-by-n complex matrix and define

$$R_i(A) = \sum_{j=1}^{n} |a_{ij}|, \qquad i = 1, \ldots, n.$$

It is well known that $|\det A| \le \prod_{i=1}^{n} R_i(A)$ and that this bound is tight, given only the $R_i(A)$. This bound is often associated with Gersgorin and it may be proved in several relatively simple ways. Suppose the numbers $r_i = R_i(A)$, $i = 1, \ldots, n$, are given, and define Γ to be the set of *all* n-by-n $B = (b_{ij})$ such that $R_i(B) = r_i$, $i = 1, \ldots, n$. The set Γ is then compact, and $A \in \Gamma$. The function $|\det B|$ is continuous. Thus, the value of

$$\max_{B \in \Gamma} |\det B|$$

is attained by some matrix in Γ (in fact one with real entries and nonnegative determinant) and is an upper bound for $|\det A|$, as $A \in \Gamma$. Consider the ith

row of a maximizing matrix with real entries and positive determinant. If the
i, j minor is a maximum in absolute value among all minors complementary to
entries in row i, then, by Laplace expansion, the ith row of our maximizing
matrix may be replaced by a row of 0's except for $\pm r_i$ in position j (+ if the i, j
minor is positive, − if it is negative) without decreasing the value of det. Ties for
maximum in the i, j minor are inconsequential. Application of this observation
to each row in turn reveals that there is a maximizing matrix with at most one
nonzero entry in each row. (There may be others.) This is an example of what
we refer to as the *concentration principal*. The resulting combinatorial problem is
easily seen to be solved by a matrix whose nonzero entries occur on a generalized
diagonal (otherwise the determinant would be 0); such a matrix has an $|\det|$ of
$\prod_{i=1}^{n} r_i$, yielding the asserted bound.

In this case, the resulting combinatorial problem is especially simple, and the
optimization renders the bound apparent.

EXAMPLE 2. In this case a finite combinatorial problem again results, but it
is most difficult.

Hadamard posed the now classical problem of finding the maximum (absolute)
value of the determinant for n-by-n matrices $A = (a_{ij})$ with $|a_{ij}| \le 1$. There
is a real and complex version of this problem; we consider the real version for
simplicity. In our terms we wish to bound $|\det A|$ given that the largest absolute
entry of A is 1. The solution to

$$\max_{|b_{ij}| \le 1} |\det B|$$

provides the best bound based upon the given information. In this case, the
constraint set $\Gamma = \{B = (b_{ij}) : |b_{ij}| \le 1\}$ is again compact, and the objective
function $|\det B|$ (or, equivalently, $\det B$) is continuous. If the (signed) i, j minor
is positive (respectively, negative) then the i, j entry of a maximizing matrix
should be +1 (respectively, −1). If the minor is 0, the entry may be chosen ± 1.
Thus the concentration principal indicates that there is an extremal maximizing
matrix, i.e., one each of whose entries is ± 1. However, although this is again
a finite set of possibilities, the resulting combinatorial problem (how to arrange
a matrix of ± 1's to maximize the determinant) has proven most difficult. The
Hadamard bound $n^{n/2}$ cannot be attained if n is not 1, 2, or a multiple of 4,
and the unresolved question of attainment for multiples of 4 has occupied great
attention in the subject of combinatorics. For n not a multiple of 4, relatively
less is known about the best bound [**BC**].

We may now summarize our general approach as follows. Given a parameter
p to be bounded for a matrix A, identify the information about A to be used. Let
Γ be the set of matrices indistinguishable from A based upon this information.
Then

$$\max_{B \in \Gamma} p(B)$$

is the best possible upper bound for $p(A)$ based upon this information. (Replace
"max" by "min" for the lower bound). If p depends continuously upon B and

if Γ is compact, then there will be an optimizing matrix occurring in Γ. (These conditions are likely to occur in most situations in which it is natural to expect a bound.) Analyze the set of optimizing matrices to see if there is a natural subset with a special combinatorial structure (perhaps extreme points of Γ). If so, inspect this special set, if possible, for the optimal value of p.

Additional examples of this approach illustrate its utility. When the combinatorial optimization problem is more subtle than in Example 1, but more feasible than in Example 2, the approach yields its most intriguing results.

EXAMPLE 3. For a real n-by-n matrix $A = (a_{ij})$, define

$$R_i^+(A) = \sum_{j\,:\,a_{ij}>0} a_{ij} \quad \text{and} \quad R_i^-(A) = - \sum_{j\,:\,a_{ij}<0} a_{ij}, \qquad i = 1,\dots,n.$$

Note that $R_i^+(A) + R_i^-(A) = R_i(A)$.

In [JN] it was shown that

$$|\det A| < \prod_{i=1}^{n} \max\{R_i^+(A), R_i^-(A)\} - \prod_{i=1}^{n} \min\{R_i^+(A), R_i^-(A)\}.$$

This improved a previous result (in which the subtracted term was absent), which, in turn, had bettered (for real matrices) the Gersgorin bound mentioned in Example 1 (based, of course, on weaker information). The proof of the above inequality may be sketched as follows. Let $r_i^+ = R_i^+(A)$ and $r_i^- = R_i^-(A)$, $i = 1,\dots,n$, and define Γ to be the set of all n-by-n matrices B with $R_i^+(B) = r_i^+$ and $R_i^-(B) = r_i^-$. Then Γ is compact and, again, $|\det A|$ is continuous; so, there exist maximizing matrices in Γ. The concentration principal then reveals that there are maximizing matrices with at most two nonzero entries in each row (one ≥ 0, one ≤ 0). A combinatorial argument further shows that a maximizing arrangement is

$$\begin{bmatrix} * & * & & & \\ & * & * & & \\ & & \ddots & \ddots & \\ & & & & * \\ * & & & & * \end{bmatrix},$$

in which the ith diagonal entry is $\max\{r_i^+, r_i^-\}$. A calculation now verifies the asserted bound.

This bound has the interesting feature that it implies that a matrix with 0 row sums is singular—an obvious fact, but a consequence of no other bound. The bound also has strong implications for M-matrices [JN, JM].

We mention two additional bounds which may be proven via the approach of this section.

EXAMPLE 4. The n-by-n matrix $A = (a_{ij})$ is an H-matrix if some diagonal multiple of A is strictly diagonally dominant. The Hadamard inequality (9.1) holds also for M-matrices and generalizes to H-matrices as

$$|\det A| \leq 2^s \left| \prod_{i=1}^{n} a_{ii} \right|$$

in which s is the greatest integer in $n/2$. This and related results may be found in [**JMP**].

EXAMPLE 5. For an n-by-n matrix $A = (a_{ij})$, define

$$s_i = \tfrac{1}{2}(|a_{i1}| + |a_{i1} - a_{i2}| + |a_{i2} - a_{i3}| + \cdots + |a_{i,n-1} - a_{in}| + |a_{in}|), \qquad i = 1, 2, \ldots, n.$$

If A is real, it is shown in [**BP**] that $|\det A| \leq \prod s_i$. Note that this implies the remarkable fact that

$$\left| \det \begin{bmatrix} 0 & 0 & \cdots & 0 & 1 & \cdots & 1 & 0 & 0 & \cdots & 0 \\ 0 & & & 0 & 0 & 1 & \cdots & 1 & 0 & \cdots & 0 \\ 1 & \ddots & & & 1 & 0 & \cdots & 0 & 0 & \cdots & 0 \\ 0 & \cdots & 0 & 1 & \cdots & 1 & 0 & 0 & 0 & \cdots & 0 \end{bmatrix} \right| \leq 1,$$

i.e., that the determinant of a $0, 1$ matrix with exactly one contiguous stretch of 1's in each row is either 0 or ± 1. Multiplication of alternate columns by -1 implies a variation of the above inequality, namely

$$|\det A| \leq \prod t_i$$

in which

$$t_i = \tfrac{1}{2}(|a_{i1}| + |a_{i1} + a_{i2}| + |a_{i2} + a_{i3}| + \cdots + |a_{i,n-1} + a_{in}| + |a_{in}|).$$

C. Matrix completion problems. In this section we discuss some very recent results about matrix completions which have classical roots. These have been of some interest in the applied functional analysis context.

By a *partial matrix* we mean a rectangular array in which some entries are *specified* (i.e., known complex numbers) and the others are *unspecified* (thought of as free complex variables). In some contexts, limited information may be known about some of the unspecified entries, but that will not be the case with the problems to be discussed here. An example of a partial matrix is

$$\begin{bmatrix} 3 & ? & i \\ ? & -2 & 0 \end{bmatrix},$$

in which the ?'s indicate unspecified entries. A *completion* of a partial matrix is simply a specification of the unspecified entries (resulting in a conventional matrix). Usually the question will be whether or not there exists a completion of a special type (such as Hermitian, positive definite, positive definite Toeplitz, a contraction, etc.). Often the existence of such a completion will imply certain obvious necessary conditions on the partial matrix. By a *partial Hermitian matrix* $A = (a_{ij})$ we mean a square partial matrix whose specified diagonal

entries are real and such that if a_{ij} is specified then so is a_{ji}, with $a_{ji} = \bar{a}_{ij}$.
By a *partial positive definite matrix* we mean a partial Hermitian matrix each
of whose specified principal submatrices is positive definite. (By a specified
portion of a partial matrix we always mean one based entirely upon specified
entries.) *Partial Toeplitz, partial positive definite Toeplitz,* and *partial positive
semidefinite* matrices, etc., are defined similarly. A rectangular matrix A such
that each eigenvalue of A^*A (or equivalently each singular value of A) is ≤ 1
is called a *contraction*. A rectangular partial matrix, each of whose specified
(possibly rectangular) submatrices is a contraction, is called a *partial contraction*.

We are interested here in two questions: (I) when a partial positive (semi)-
definite matrix has a positive (semi)definite completion; and (II) when a partial
contraction has a completion which is a contraction. In each case the assumption about the partial matrix (partial positive definite or partial contraction) is
necessary for the existence of a completion of the stated sort. The questions (I)
and (II) are closely related.

We first consider positive (semi)definite completions.

It is a classical analysis result of Carathéodory/Fejer (not usually stated in
these terms) that a partial positive definite (n-by-n) Toeplitz matrix in which
the first $k < n$ bands (from the diagonal out) are specified and the remaining
bands are unspecified may be completed to a positive definite Toeplitz matrix.
A very interesting recent paper [**DGo**] shows that the Toeplitz structure in the
Carathéodory/Fejer theorem is not crucial. We call a square partial matrix
banded if all the entries within some bandwidth (symmetric from the diagonal)
are specified and all entries outside are unspecified. Recall that a conventional
matrix $A = (a_{ij})$ is called banded with bandwidth k if $a_{ij} = 0$ whenever $|i-j| >
k$. We adopt the same convention for measuring the bandwidth of a banded
partial matrix by numbering the main diagonal as the 0th band.

Assuming that A is an n-by-n banded partial positive definite matrix of bandwidth $k \geq 0$, three principal conclusions are drawn in [**DGo**]:

(i)   there is a positive definite completion of A,

(ii)   among the convex set of positive definite completions of A there
        is a unique one of maximum determinant;

and

(iii)   the determinant maximizing positive definite completion of A is
        the unique nonsingular completion whose inverse is banded (in
        the usual sense) with bandwidth k.

The authors also gave a version of these results for "block banded" partial positive definite matrices. Conclusion (i) may be viewed as indicating a set of
patterns (for the specified entries) for which *all* partial positive definite matrices
may be completed to positive definite matrices.

Though not noted by the authors of [**DGo**], conclusions (i), (ii), and (iii) may also be viewed as a generalization of Hadamard's inequality (9.1) in the following way. Consider a conventional positive definite matrix A and then "free up" all off-diagonal entries and consider all positive definite completions of the resulting "diagonal" partial Hermitian matrix. According to (iii), the determinant maximizing positive definite completion has diagonal inverse and, therefore, is diagonal; its determinant is thus the product of its diagonal entries. Since the original conventional matrix A is one of the positive definite completions, its determinant must be $\leq$ its diagonal product, with equality exactly when A is diagonal, i.e., Hadamard's inequality. This connection with inequalities is of natural fundamental interest in matrix theory.

It is natural to ask for all patterns (of the specified entries) such that a partial positive definite matrix necessarily has a positive definite completion. (By [**DGo**] the banded patterns are among these.) In this regard the positive semidefinite and positive definite versions of question (I) are the same. The simplest example which indicates that not all patterns insure positive (semi)definite completions is the partial matrix (with unspecified entries determined by x_1 and x_2):

$$\begin{bmatrix} 1 & 1 & x_1 & -1 \\ 1 & 1 & 1 & x_2 \\ \overline{x}_1 & 1 & 1 & 1 \\ -1 & \overline{x}_2 & 1 & 1 \end{bmatrix}.$$

It is partial positive semidefinite as the only specified principal submatrices are

$$[1], \qquad \begin{bmatrix} 1 & 1 \\ 1 & 1 \end{bmatrix}, \quad \text{and} \quad \begin{bmatrix} 1 & -1 \\ -1 & 1 \end{bmatrix}.$$

However, x_2 completes both partial principal submatrices

$$\begin{bmatrix} 1 & 1 & x_2 \\ 1 & 1 & 1 \\ \overline{x}_2 & 1 & 1 \end{bmatrix}, \quad \text{rows and columns } 2, 3, 4,$$

and

$$\begin{bmatrix} 1 & 1 & -1 \\ 1 & 1 & x_2 \\ -1 & \overline{x}_2 & 1 \end{bmatrix}, \quad \text{rows and columns } 1, 2, 4.$$

The former requires $x_2 = 1$ for positive semidefinite completion and the latter requires $x_2 = -1$ for positive semidefinite completion. As these are in conflict and all principal submatrices of a positive semidefinite matrix must be positive semidefinite, no positive semidefinite completion of the 4-by-4 partial matrix is possible. The characterization of completable patterns has been addressed in [**GJSW1**], and the most natural way of describing patterns is the undirected graph $G = G(A)$ of the specified entries. The graph G has vertices $1, 2, \ldots, n$ if the partial Hermitian matrix A is n-by-n, and an edge between vertices i and j

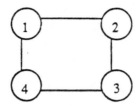

FIGURE 9.1

if and only if the i,j entry of A is specified. Thus, the example above has graph as in Figure 9.1.

Undirected graphs are appropriate because we assume the partial matrix is Hermitian. For completeness, we informally and briefly mention some standard notions from graph theory. A *path* is just a sequence: $\{i_1, i_2\}, \{i_2, i_3\}, \ldots,$ $\{i_{k-1}, i_k\}$ of concatenated edges, and a *circuit* is a path for which $i_k = i_1$, i.e., which begins and ends at the same vertex. A *simple circuit* has no self-intersections, i.e., $i_1, i_2, \ldots, i_{k-1}$ are all different and a minimal (simple) *circuit* is one no proper subset of whose vertices are themselves the vertices of a circuit. A *chord* of a circuit is an edge which joins two vertices of the circuit which are not adjacent in the circuit.

The key notion which allows simple description of completable patterns is that of a *chordal graph*: we call G chordal if it has no minimal simple circuits of 4 or more edges, i.e., any circuit of 4 or more edges must have a chord. The graph of the above example is the simplest nonchordal graph. Chordal graphs have previously arisen in matrix analysis, in the study of optimal Gaussian elimination for large sparse matrices; a good general reference including supporting graph theory is [**Go**]. Another name often used for chordal graphs is triangulated; the motivation is clear.

We note that in the remaining discussion of positive definite completions we assume without loss of generality that all diagonal entries of a partial positive definite matrix are specified; as any unspecified ones could be chosen arbitrarily large, the question of a positive definite completion then rests upon that of a partial principal submatrix all of whose diagonal entries are specified. The principal result of [**GJSW1**] is

THEOREM 9.1. *Every partial positive definite matrix with graph G has a positive definite completion if and only if G is chordal.*

It is worth noting that chordality is more general than (permutation similarity to) banded or block banded. For example, the tree shown in Figure 9.2 is, of course, chordal, while no numbering of the vertices yields even a block banded configuration. This theorem solves one (limited) subclass of problems of the type (P1) as we indicated toward the end of Chapter 8.

Proof of the above theorem is easily summarized, though complete details are somewhat lengthy. The necessity of chordality rests upon exhibition of a

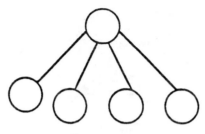

FIGURE 9.2.

general class of counterexamples. Since every principal submatrix of a positive (semi)definite matrix must be positive (semi)definite, it suffices to show by example that a single simple circuit of 4 or more edges is itself not completable, as any nonchordal graph must have such a subgraph. Specification of the data for such a pattern as

$$\begin{bmatrix} 1 & 1 & & & & -1 \\ 1 & 1 & 1 & & ? & \\ & 1 & 1 & 1 & & \\ & ? & \ddots & \ddots & \ddots & \\ & & & & & 1 \\ -1 & & & & 1 & 1 \end{bmatrix}$$

provides the necessary partial positive definite matrices which have no positive definite completions.

The sufficiency of chordality rests on the following combinatorial observation. The edges *missing* (relative to the complete graph) from a chordal graph may be ordered so that when they are added one at a time, each successive graph is chordal. If an unspecified entry is specified in association with the addition of each of these edges in turn, it turns out that a unique maximal partial principal submatrix of our partial Hermitian matrix is completed, and it suffices to exhibit a completion with positive determinant. In fact, without loss of generality, the maximal partial principal submatrix may be taken to be of the form

$$A(\mathbf{z}) = \begin{bmatrix} a_{11} & a_{12}^* & \mathbf{z} \\ a_{12} & A_{22} & a_{23} \\ \mathbf{z} & a_{23}^* & a_{33} \end{bmatrix}$$

with $\mathbf{z}$ the unspecified entry. Since the upper left principal submatrix

$$\begin{bmatrix} a_{11} & a_{12}^* \\ a_{12} & A_{22} \end{bmatrix}$$

is specified, all leading principal minors of $A(\mathbf{z})$ save the last are positive, so that a choice of $\mathbf{z}_0$ with $\det A(\mathbf{z}_0)$ positive means that $A(\mathbf{z}_0)$ is positive definite.

But, this may be deduced from the banded result of [**DGo**] or by analyzing $\det A(\mathbf{z}) = -(\det A_{22})\mathbf{z}\bar{\mathbf{z}} + 2\operatorname{Re}(\alpha\mathbf{z}) + \det A(0)$, in which

$$\alpha = \pm \det \begin{bmatrix} a_{12} & A_{22} \\ 0 & a_{23} \end{bmatrix}.$$

We note that there is a real solution $\mathbf{z}$ if the data is real. A positive definite completion may then be constructed inductively from the chordal ordering of the missing edges mentioned above.

This theorem gives a complete generalization of conclusion (i) above from [**DGo**]; it also describes a large class of affine subspaces of matrix space which intersect the cone of positive definite matrices.

Conclusions (ii) and (iii) of [**DGo**] also generalize completely as necessary conditions for the existence of a positive definite completion of a partial positive definite matrix, *irrespective* of the pattern of unspecified entries. This exhibits another nice use of optimization in matrix analysis. We continue to assume that all diagonal entries of a partial positive definite matrix are specified. It is also shown in [**GJSW1**] that

THEOREM 9.2. *If A is a partial positive definite matrix with a positive definite completion, then*

(ii′)   *there is a unique positive definite completion of A with maximum determinant; and*

(iii′)   *this unique determinant maximizing completion is the unique one whose inverse has entries equal to 0 in all positions corresponding to unspecified entries of A.*

It is clear that (ii) and (iii) are the special cases of (ii′) and (iii′) in which A is banded. Again we summarize the proof. First consider the compact set S of positive semidefinite completions of A. Since the determinant function is strictly log concave on the positive semidefinite matrices [**HJ**], it attains a unique maximum on S. This maximum must be attained at a positive definite completion of A (or the determinant would be 0), i.e., in the interior of S; this verifies (ii′). Consider $\det A$ as a function of its unspecified entries. Since our interior maximum must be the unique critical point of $\det A$ in the interior of S (because of strict log concavity), it suffices to note that the partial of $\det A$ with respect to the real part (imaginary part) of the i,j entry is a constant multiple of the real part (imaginary part) of its complementary minor. But, by the cofactor form of the inverse, this complementary minor is the numerator of the j,i entry of the inverse. Thus, all entries of the inverse, corresponding to unspecified entries of A, are 0 exactly at a critical point, which verifies (iii′).

By combining it with a class of simple determinantal formulae for A based upon the 0 pattern of A^{-1} [**BJ1**], implications of the above theorem for determinantal inequalities for positive definite matrices have been studied in depth in

[**BJ2**]. These are similar in spirit to the way in which it was noted above that the observations of [**DGo**] generalize Hadamard's inequality (9.1), and, in fact, the inequalities so produced are all possible generalizations of (9.3) of a ratio type.

The results mentioned thus far have all dealt with positive definite completions, in response to question (I). The first theorem might be viewed slightly differently. It implies that a partial Hermitian matrix with chordal graph may be completed so that the minimum eigenvalue of the completion is the same as the smallest of the minimum eigenvalues of all specified principal submatrices. Recall that the inertia of an Hermitian matrix A is just a summary

$$i(A) = (i_+(A), i_-(A), i_0(A))$$

of the number of positive (i_+), negative (i_-), and zero (i_0) eigenvalues of A, counting multiplicity. Our alternate view of the first theorem suggests the following variation upon question I: when may a general partial Hermitian matrix be completed so as to have as general an inertia as the interlacing inequalities [**HJ**] applied to its specified principal submatrices will allow? This question is dealt with in [**JR1**], and again chordality plays a key role. Although certain technical assumptions are necessary (nonsingularity of certain strategically placed specified principal submatrices), inertia of completions is constrained only by interlacing if and only if the graph G is chordal. The methodology is, however, more basic as the luxury of positive specified principal minors is not present.

We finally turn to the very recent resolution of question (II) in [**JR2**] from a combinatorial point of view. A number of references which relate (II) to other matters may be found there.

Question (II) is linked to (I) by the following elementary observation. If A is an m-by-n complex matrix, then A is a contraction if and only if the $(m+n)$-by-$(m+n)$ matrix $\left[\begin{smallmatrix} I & A \\ A^* & I \end{smallmatrix}\right]$ is positive semidefinite. We call an n-by-m partial matrix *decomposable* if there exist permutation matrices Q_1 (n-by-n) and Q_2 (m-by-m) such that

$$Q_1 A Q_2 = \begin{bmatrix} B_{11} & B_{12} \\ B_{21} & B_{22} \end{bmatrix}$$

in which the blocks B_{12} and B_{21} consist entirely of unspecified entries. (Here, blocks B_{11} and B_{22} may be rectangular, and one of the parts $[B_{21} B_{22}]$ or $\left[\begin{smallmatrix} B_{12} \\ B_{22} \end{smallmatrix}\right]$, but not both, may be empty.) Otherwise, A is called indecomposable. Since the decomposable case of question (II) depends only upon its "diagonal" blocks, it suffices to consider the indecomposable case. The result in response to (II) is then

THEOREM 9.3. *For an m-by-n indecomposable pattern P for the specified entries, the following are equivalent:*

(a)   *every m-by-n partial contraction with the pattern P for its specified entries can be completed to an m-by-n contraction;*

(b)   *the pattern P has no subpattern (lying in the intersection of 2 rows and 2 columns) of the form*

$$\begin{bmatrix} ? & \\ & ? \end{bmatrix} \quad \text{or} \quad \begin{bmatrix} & ? \\ ? & \end{bmatrix}$$

in which the blanks (resp. question marks) denote specified (resp. unspecified) entries; and

(c)   *P is permutation equivalent to*

$$\begin{bmatrix} B_{11} & B_{12} & \cdots & B_{1r} \\ B_{21} & B_{22} & & \\ \vdots & & \ddots & \\ B_{r1} & & & B_{rr} \end{bmatrix}$$

in which the blocks B_{ij} (which may be rectangular) are fully specified for $i \geq j$ and fully unspecified for $i < j$, i.e., the permuted pattern is "block lower triangular."

That (a) is implied by (b) or (c), which are equivalent to each other on combinatorial grounds, is implied by the positive definite completion result for chordal graphs, via the elementary observation above. The reverse implication again requires a specially constructed class of examples, as the positive definite chordal theorem does not match. An indicative example is

$$\frac{1}{\sqrt{2}} \begin{bmatrix} 1 & ? & 1 \\ ? & 1 & 1 \end{bmatrix}$$

which is an indecomposable partial contraction without a contraction completion. By inspecting the rows, each ? must be completed by 0 for an opportunity at a contraction, but

$$\frac{1}{\sqrt{2}} \begin{bmatrix} 1 & 0 & 1 \\ 0 & 1 & 1 \end{bmatrix}$$

is not a contraction.

There are a number of interesting further questions in the general area of matrix completions.

For example which sequences of specified bands, $0, i_1, i_2, \ldots, i_k$, insure that a partial positive definite Toeplitz matrix has a positive definite Toeplitz completion? The Carathéodory/Fejer theorem implies that the sequence $0, 1, 2, \ldots, k$

works and it may be shown that a sequence of the form $0, p, 2p, \ldots, kp$ does also. Leiba Rodman and I conjecture that these are all.

In the general (not necessarily chordal) partial Hermitian matrix case, it would be of interest to know what governs the disparity between the largest which the smallest eigenvalue of a completion may be made and the smallest of all the minimum eigenvalues of specified principal submatrices. The former is no more than the latter (and equal in the chordal case). An answer in terms of the positions of specified entries (perhaps a disparity from chordality) would be most intriguing.

10. The Spectrum of a Matrix with Respect to an Algebra

This chapter addresses a very different class of matrix approximation problems. The issue is how does one determine if a matrix w in an intersection set

$$\mathbf{S} \cap \Delta_Y$$

lies deeply in the interior of the set or near the boundary. Here various norms could be used to measure distance. This essentially is (P2) stated in the introduction to Part II. We shall focus on a very special case of a more general problem.

This chapter describes how the problem arises in engineering, describes an elegant reformulation, and lists a few attempts at a compromise solution. Current results are not mathematically definitive, probably because definitive results are impossible. We hope the reader will find some of the mathematical questions here interesting.

Suppose that we are a step further along in the narrowband design sketched near Figure 7.1. Say we have found a z so that a performance specification

$$(10.1) \qquad\qquad \Gamma(\omega_0, z) \le c$$

is met. Let us reexamine the assumptions behind the setup. We assume that part of the system was given to us and that we know it precisely (see Figure 10.1). It is more realistic to assume that the given system is not known perfectly but that there is some error in our understanding. We can actually model the error as a matrix parameter δ with norm smaller than a given ε as in Figure 10.2.

Clearly, the performance of the system now is a function $\Gamma(\omega, z, \delta)$ of not just ω and z but of which value of δ is "really true." Meeting the specifications means

$$(10.2) \qquad\qquad \Gamma(\omega, z, \delta) \le c$$

but since we don't know the "true" δ, this must hold for all $|\delta| \le \varepsilon$ and δ in a subspace $\mathbf{S}$ of admissible perturbations. Henceforth assume we have normalized δ so that $\varepsilon = 1$.

Let us now be more specific about the form of the function Γ. Fix z and fix ω_0. Then Figure 10.2 is equivalent to Figure 10.3 and so the response function

FIGURE 10.1.

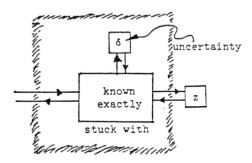

FIGURE 10.2.

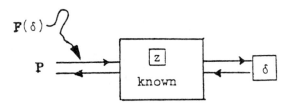

FIGURE 10.3.

for the total system is $\mathbf{F}(\delta)$ where $\mathbf{F}$ is a L.F.T. If as before (in Chapter 7) the performance measure has the form $\|p\|^2$ or $\|p - e\|^2$ etc., then the specification (10.1) becomes $\|\mathbf{F}(\delta)\| \le c$. Consequently the key problem has the form

(PROB)   *Given* $\mathbf{F}$ *an L.F.T., is* $\|\mathbf{F}(\delta)\| \le 1$ *for all* δ *in* $\mathbf{S}$ *with* $\|\delta\| \le 1$?

Lemma 8.3 tells us that [PROB] is equivalent to or approximable by

(PROB')   *Is* $\mathbf{S} \cap \overline{\mathbf{B}}M_{m \times n}$ *contained in the disk* Δ_Y?

which is a special case of problem (P2) which began Part III.
    Note that a disk

$$\Delta_Y = \left\{ r \colon \left(Y \begin{pmatrix} r \\ I \end{pmatrix} x, \begin{pmatrix} r \\ I \end{pmatrix} x \right) \le 0 \text{ all } x \right\},$$

because of the Cholesky decomposition $Y = LDL^*$, can be rewritten

$$(j_1[\alpha r + \beta]x, [\alpha r + \beta]x) \le -(j_2 \gamma x, \gamma x),$$
$$[\alpha r + \beta]^* j_1 [\alpha r + \beta] \le -j_2,$$

where $L = \begin{pmatrix} \alpha^* & 0 \\ \beta^* & \gamma^* \end{pmatrix}$ and $D = \begin{pmatrix} j_1 & 0 \\ 0 & j_2 \end{pmatrix}$ with j_1 and j_2 diagonal, with only ± 1 or 0 on the diagonal. We call Δ_Y an *interior* disk if both $j_1 \geq 0$ and $-j_2 \geq 0$. Note that interior disks are convex sets. Another formulation of (PROB) is

(PROB'')   *Do all δ in $\mathbf{S} \cap \overline{\mathbf{B}} M_{m \times n}$ satisfy $[\alpha\delta + \beta]^* j_1 [\alpha\delta + \beta] \leq -j_2$?*

A cultural note gives perspective. In control design the subject of δ's and ε's is called *plant uncertainty*. Accounting for plant uncertainty has been discussed by engineers for some years, e.g., Horowitz in the 1960s; however, due largely to efforts of John Doyle and Gunther Stein, it is now widely believed that a useful mathematical formulation of a practical design problem *must* take plant uncertainty into account. Mathematical methodologies which fail to account for it in a serious way do not lead to practical designs. A different set of remarks is that the problem currently arises in a context where one only needs to answer (PROB) for 3×3 to 20×20 matrices; so, computation is eminently feasible. Unfortunately such a computation must be done many, many times, for example, at each frequency ω on a large grid of ω_k's.

An engineer has substantial discretion in how he formulates this problem mathematically. If one is presented an $\mathbf{S}$ which is "complicated" it is possible to map the problem to a higher dimensional space and get a simple $\mathbf{S}$. For example there is a standard engineering construction which always allows one to take $\mathbf{S} = $ the diagonal matrices but in a very high dimensional space. Doyle always elects to take $\mathbf{S} = $ the diagonal matrices even though the space in which everything lies is large.

The main thing we shall do in this chapter is describe some elegant (approximate) reformulations of (PROB) which actually lead to the computational methods which Doyle and the Honeywell group use for aircraft control design. In tone this chapter is different than that of the earlier ones in that everything we do in it is a *compromise*.

First Doyle defines the *spectrum of a matrix* W with respect to a set $\mathbf{Q}$ of matrices as

$$\sigma_{\mathbf{Q}}(W) = \{\xi \in \mathbf{Q} : W - \xi \text{ is not invertible}\}.$$

Define the *spectral radius* $\mu_{\mathbf{Q}}(W)$ of W w.r.t. $\mathbf{Q}$ by

$$\mu_{\mathbf{Q}}(W) = \sup_{\xi \in \sigma_{\mathbf{Q}}(W)} \inf_x \frac{\|\xi x\|}{\|x\|}.$$

A little reflection convinces one that this is the natural definition for spectral radius when matrices are involved. Doyle tentatively calls this spectral radius the *structured singular value* (SSV).

EXAMPLE 1. $\mathbf{Q} = \{\lambda I : \lambda \in \mathbb{C}\}$. Then $\sigma_{\mathbf{Q}}(W) = \sigma(W)$ and $\mu_{\mathbf{Q}}(W) = $ spec. rad.

EXAMPLE 2. $\mathbf{Q}=$ all $n \times n$ matrices. We shall see that $\mu_{\mathbf{Q}}(W) = \|W\| =$ highest singular value of W.

While these definitions hold for very general $\mathbf{Q}$ we shall prove most theorems when $\mathbf{Q} =$ the block diagonal matrices. Restricting $\mathbf{S}$ to equal the diagonal matrices is physically justified since in Figure 10.2 we can take δ to be a diagonal matrix (with as many parameters as determine the exactly known part of the box). The relationship between (PROB) and structured singular values is

THEOREM 10.1 (DOYLE, STEIN, WALL [DSW]). *Given* (PROB) *one can write down a block* 2×2 *matrix* W *such that the answer is yes if* $\mu_{\mathbf{Q}}(W) \leq 1$ *where* $\mathbf{Q} = \left(\begin{smallmatrix} M_{k\times k} & 0 \\ 0 & dg \end{smallmatrix} \right)$. *Here we assume that each* $F(\delta)$ *is a* $k \times k$ *matrix. If* $F(\delta) = a + b\delta(1 - d\delta)^{-1}c$ *with* $\|d\| < 1$, *then* $\mu_{\mathbf{Q}}(W) \leq 1$ *is equivalent to* (PROB) *having answer yes. An explicit formula is* $W = \left(\begin{smallmatrix} -a & b \\ -c & d \end{smallmatrix} \right)$.

The implication of the theorem is that Doyle forgets (PROB) and turns his full attention to computing the SSV of a matrix W. The proof which follows simply constructs W. We need a lemma which we present in excess generality.

LEMMA 10.2 (DOYLE). *If* $\mathbf{Q} \subset M_{n\times n}$ *is closed under scalar multiplication, under inverses when they exist, and contains the identity, then*

$$\mu_{\mathbf{Q}}(W) = \sup_{\beta \in \mathbf{BQ}} \text{ spec. rad.}(\beta W).$$

PROOF. If ξ is invertible, then

$$\inf_x \frac{\|\xi x\|}{\|x\|} = \inf_y \frac{\|y\|}{\|\xi^{-1}y\|} = \left[\sup_y \frac{\|\xi^{-1}y\|}{\|y\|} \right]^{-1} = \|\xi^{-1}\|^{-1}.$$

Also $W - \xi = \xi(\xi^{-1}W - I)$ is not invertible if and only if $I - \xi^{-1}W$ is not invertible. Thus

$$\mu(W) = \sup\{\|\xi^{-1}\|^{-1} : I - \xi^{-1}W \text{ is not invertible and } \xi \in \mathbf{Q}\}$$
$$= \sup\{\|\xi^{-1}\|^{-1} : 1 \in \sigma(\xi^{-1}W) \text{ and } \xi \in \mathbf{Q}\}.$$

Since ξ^{-1} is in $\mathbf{Q}$ when

(10.3)
$$\xi \leq \sup\{\|\alpha\|^{-1} : \|\alpha\|^{-1} \in \sigma(\|\alpha\|^{-1}\alpha W) \text{ and } \mathbf{Q}\}$$
$$= \sup\{r : r \in \sigma(\beta W) \text{ and } \beta \in \mathbf{Q} \text{ with } \|\beta\| = 1\}.$$

Because the identity is in $\mathbf{Q}$, inverses of elements of $\mathbf{Q}$ are dense in $\mathbf{Q}$ which implies that the inequality in (10.3) also goes the other way and so is an equality.

PROOF[1] OF THEOREM 10.1. Let $\begin{pmatrix} a & b \\ c & d \end{pmatrix}$ be the coefficient matrix of $\mathbf{F}$. That is

$$\mathbf{F}(\delta) = a + b\delta(1 - d\delta)^{-1}c$$

with each $\mathbf{F}(\delta)$ a $k \times k$ matrix. Set

$$(10.4) \qquad\qquad W = \begin{pmatrix} -a & b \\ -c & d \end{pmatrix}.$$

Assume for a while that $1 - d\delta$ is invertible. Then $I - W\begin{pmatrix} \eta & 0 \\ 0 & \delta \end{pmatrix} = \begin{pmatrix} 1+a\eta & -b\delta \\ c\eta & 1-d\delta \end{pmatrix}$ has an inverse which (cf. [**BGK**], (1.9)) by block Gauss elimination is

$$\begin{pmatrix} T^{-1} & T^{-1}b(1-d\delta)^{-1} \\ -(1-d\delta)^{-1}cT^{-1} & -(1-d\delta)^{-1}cT^{-1}b(1-d\delta)^{-1} + (1-d\delta)^{-1} \end{pmatrix}$$

with $T = 1 + a\eta + b\delta(1-d\delta)^{-1}c\eta = 1 + \mathbf{F}(\delta)\eta$. The inverse exists if and only if T is invertible which happens for all $\|\eta\| < 1$ if and only if $\|\mathbf{F}(\delta)\| \leq 1$. We have

If $(1 - d\delta)$ is invertible, then $\left(I - W\begin{pmatrix} \eta & 0 \\ 0 & \delta \end{pmatrix}\right)$ is invertible for all $\|\eta\| < 1$ if and only if $\|\mathbf{F}(\delta)\| \leq 1$.

Now $\mu_{\mathbf{Q}}(W) \leq 1$ means that $\left(I - W\begin{pmatrix} \eta & 0 \\ 0 & \delta \end{pmatrix}\right)$ is invertible for all $\|\eta\|, \|\delta\| < 1$. Consequently, if $\mu_{\mathbf{Q}}(W) \leq 1$, then $\|\mathbf{F}(\delta)\| \leq 1$ for all $\{\delta : \|\delta\| < 1$ except possibly when $1 \in \sigma(\delta d)\}$. Now $\mathbf{F}$ is defined at singularities of $(1 - d\delta)$ by taking limits. This gives $\|\mathbf{F}(\delta)\| \leq 1$ even for $1 \in \sigma(\delta d)$.

Conversely, if $\|\mathbf{F}(\delta)\| \leq 1$ for all $\|\delta\| < 1$, then by formula (10.5)

$$\begin{pmatrix} I & 0 \\ 0 & b \end{pmatrix} \left[I - W\begin{pmatrix} \eta & 0 \\ 0 & \delta \end{pmatrix} \right]^{-1} \begin{pmatrix} I & 0 \\ 0 & c \end{pmatrix}$$

is not singular for any $\|\eta\|, \|\delta\| < 1$. Note that if $\|d\| < 1$, then $\mu_{\mathbf{Q}}(W) \leq 1$ if and only if $\|\mathbf{F}(\delta)\| \leq 1$ for all $\|\delta\| \leq 1$.   Q.E.D.

All of these motivational engineering allusions may be confusing some simple issues. For example, if Δ_Y is an interior disk, then the problem

$$(10.5) \qquad Is \ \{\delta : \delta \text{ is diagonal and } \|\delta\| \leq 1\} \text{ contained in } \Delta_Y?$$

will be equivalent to a SSV problem via Theorem 10.1. This is because

$$\Delta_Y = \{\delta : \|\mathbf{F}(\delta)\| \leq 1\}$$

for an $\mathbf{F}$ which is actually *linear*. For this $\mathbf{F}$ in Theorem 10.1, the crucial matrix $d = 0$, which implies $1 - d\delta$ is always invertible, and so the answer to (10.5) is yes if and only if $\mu_{\mathbf{Q}}(W) \leq 1$; here W has the form $\begin{pmatrix} -a & b \\ -c & 0 \end{pmatrix}$.

All of this serves as motivation for studying $\mu_{\mathbf{Q}}$ which is indeed a challenging mathematics problem. Henceforth $\mathbf{Q}$ will always be an algebra of block diagonal matrices.

---

[1]One should see [**DSW**] since the original argument is quite different.

Doyle's compromise method for computing $\mu_{\mathbf{Q}}(W)$ is to form the function

$$N(D) = \|DWD^{-1}\|$$

on the diagonal matrices D with positive entries. He then advises a designer to compute $\nu_{\mathbf{Q}}(W) \triangleq \inf_D \{N(D): D \text{ diagonal and } DAD^{-1} = A \text{ for all } A \in \mathbf{Q}\}$ using a certain gradient descent method. One can easily prove[2]

THEOREM 10.3 (DOYLE). $\mu_{\mathbf{Q}}(W) \leq \nu_{\mathbf{Q}}(W)$.

So $\nu_{\mathbf{Q}}(W)$ is an upper bound for $\mu_{\mathbf{Q}}(W)$. Doyle did computer experiments and the worst error of $\mu_{\mathbf{Q}}/\nu_{\mathbf{Q}}$ on 50,000 pseudorandom-generated matrices was about .95. The experiments were based on computing two numbers. One was the upper bound $\nu(W)$ for $\mu(W)$; the other was the lower bound based on a gradient ascent iteration to approximate $L(W)$.

THEOREM 10.4.

$$L(W) \triangleq \sup_{\substack{\delta \in \mathbf{Q} \\ \delta \text{ is unitary}}} \text{spec. rad.}(\delta W) \text{ equals } \mu_{\mathbf{Q}}(W).$$

It was these two numbers whose ratio was no worse than .95. How far apart these numbers can be for an $n \times n$ matrix is a completely open question. At the moment a W is known which produces an error ratio of .85; that is the worst known case.

PROOF OF THEOREM 10.4. The spectral radius inequality is an immediate consequence of the fact that a function of several complex variables analytic in a domain $\mathbf{D}$ takes its maximum on the distinguished boundary of $\mathbf{D}$. The distinguished boundary of $\{\delta \text{ diag.}: |\delta| \geq 1\}$ is $\{\delta \text{ diag.}: \delta \text{ unitary}\}$.

Now $\mu_{\mathbf{Q}}(W) = 1$ means $(1 - W\delta^{-1})^{-1}$ is analytic for all $\|\delta^{-1}\| < 1$ and blows up at some δ_0 with $\|\delta_0^{-1}\| = 1$. Consider the function $f(\delta) \triangleq ((1 - W\delta^{-1})^{-1}x, y)$ where x, y are vectors chosen so that this function is singular at δ_0. This function is analytic on δ in the open polydisk, and consequently the maximum of f must be taken at a point on the distinguished boundary. In particular if any singularity exists on the boundary one must exist on the distinguished boundary. (One can show this by perturbing f slightly to get a function with no singularities on the closed polydisk.)

Other items which are useful in practice are

(1)   If $W \in M_{n \times n}$ and $\mathbf{Q}$ is all block diagonal matrices with a given definition of block structure, then $\sigma_{\mathbf{Q}}(W) = \nu(W)$ if the number of blocks is less than 4. Doyle asserts this to be true in [**Doy1-Doy2**].

---

[2]PROOF. First note $\mu_{\mathbf{Q}}(W) = \mu_{\mathbf{Q}}(DWD^{-1})$ for D which commute with all A in $\mathbf{Q}$. Then use $\mu_{\mathbf{Q}}(M) \leq \|M\|$.

(2)   The function $N(e^D)$ is a convex function[3] on positive diagonal
         D, which appears in [**DSa**].

The interested reader should see Fan and Tits [**FT**] who give a very different
approach to computing the structured singular value. It is an ascent algorithm
which approaches $\mu_{\mathbf{Q}}(W)$ from below. Since $N(D)$ for each diagonal matrix is
an upper bound on $\mu_{\mathbf{Q}}(W)$, the [**FT**] approach is especially valuable in that at
each iteration it gives a lower bound on positive diagonal D.

Now we change topics.

The objective is to sketch what roughly is the method behind Doyle's design
procedure for aircraft controllers. We do not work at the full generality of the
method and we de-emphasize constraints which though physically important add
no new mathematical difficulty.

Consider the control problem in Figure 2.1, but with the wrinkle that the
plant P is not known perfectly. Recall that when P is known perfectly as in
Chapter 2 we parametrized the possible design by T the "closed loop" transfer
function. The objective was to find T in $\mathbf{A}$ and in a certain disk Δ_K^R. The disk
constraint could be reformulated if we chose by writing down a certain linear
fractional (in this case linear) map $\mathbf{L}_{U(\omega)}$ for which

$$(10.6) \qquad \|\mathbf{L}_{U(\omega)}(T(\omega))\| \leq 1 \Leftrightarrow T \in \Delta_K^R.$$

If P is not known perfectly, then the "parameter extraction" trick in Figure
10.2 allows us to write

$$(10.6') \qquad P(\omega) = \mathbf{F}_{V(\omega)}(\delta(\omega))$$

---

[3]One can be misled by statements in the literature which seem to indicate that $N(D)$ itself
is a convex function of D, for D which are positive diagonal matrices.

COUNTEREXAMPLE. Set $W = \left(\begin{smallmatrix} 0 & 1 \\ 0 & 0 \end{smallmatrix}\right)$, $X = \left(\begin{smallmatrix} 1/8 & 0 \\ 0 & 2 \end{smallmatrix}\right)$, and $Y = \left(\begin{smallmatrix} 10 & 0 \\ 0 & 6 \end{smallmatrix}\right)$. Then

$$N(X) = \left\|\begin{pmatrix} \frac{1}{8} & 0 \\ 0 & 2 \end{pmatrix}\begin{pmatrix} 0 & 1 \\ 0 & 0 \end{pmatrix}\begin{pmatrix} 8 & 0 \\ 0 & \frac{1}{2} \end{pmatrix}\right\| = \left\|\begin{pmatrix} 0 & \frac{1}{16} \\ 0 & 0 \end{pmatrix}\right\| = \frac{1}{16},$$

$$N(Y) = \left\|\begin{pmatrix} 10 & 0 \\ 0 & 6 \end{pmatrix}\begin{pmatrix} 0 & 1 \\ 0 & 0 \end{pmatrix}\begin{pmatrix} \frac{1}{10} & 0 \\ 0 & \frac{1}{6} \end{pmatrix}\right\| = \left\|\begin{pmatrix} 0 & \frac{10}{6} \\ 0 & 0 \end{pmatrix}\right\| = \frac{10}{6},$$

$$N\left(\frac{X+Y}{2}\right) = N(X+Y) = \left\|\begin{pmatrix} 10\frac{1}{8} & 0 \\ 0 & 8 \end{pmatrix}\begin{pmatrix} 0 & 1 \\ 0 & 0 \end{pmatrix}\begin{pmatrix} (10\frac{1}{8})^{-1} & 0 \\ 0 & \frac{1}{8} \end{pmatrix}\right\| = \frac{10\frac{1}{8}}{8}.$$

$$1 < \frac{10\frac{1}{8}}{8} \not\leq \left[\frac{1}{16} + \frac{10}{6}\right]\frac{1}{2} < 1$$

so

$$N\left(\frac{X+Y}{2}\right) \not\leq [N(X) + N(Y)]\frac{1}{2}.$$

REMARK. $N(D)$ *is subadditive.*

$$N(X+Y) = \|(X+Y)W(X+Y)^{-1}\|$$
$$\leq \|XW(X+Y)^{-1}\| + \|YW(X+Y)^{-1}\|.$$

Now $XW(X+Y)^{-2}W^*X^* \leq XWX^{-2}W^*$, since $(X+Y) \geq X$ implies $(X+Y)^{-2} \leq X^{-2}$.
Consequently $\|XW(X+Y)^{-1}\| \leq \|XWX^{-1}\| = N(X)$. Likewise, the second term is $\leq N(Y)$.
Consequently we have $N(X+Y) \leq N(X) + N(Y)$.

where $\delta(\omega)$ is some diagonal matrix of norm ≤ 1. The unfortunate thing is that we do not know which $\delta(\omega)$ occurs. Let $P_0(\omega) = \mathbf{F}_{V(\omega)}(0)$ and call it the nominal plant. Then we can parametrize[4] everything in terms of $T = P_0 C(1 + P_0 C)^{-1}$ which must be in $\mathbf{A}$ for the closed loop system to be stable. Since the formulas (10.6), (10.6') are linear fractional we can combine them to get that our main design problem is: Given $F_\omega(\tau, \delta)$ a function which is linear fractional in each of its two variables τ and δ,

(10.7)   *Find $T \in \mathbf{A}$ such that* $\sup\limits_{\omega} \sup\limits_{\substack{\delta \\ \text{diag} \\ \|\delta\| \leq 1}} |F_\omega(T(i\omega), \delta)| \leq 1.$

For MIMO systems one has matrix-valued T and F_ω, but the main problem has exactly the same form.

Doyle's numerical approach to the problem works like this:

(1)   Use Theorem 10.1 to obtain a matrix $W_\omega(\tau)$ so that

$$\mu_\mathbf{Q}(W_\omega(T(j\omega))) \leq 1 \Rightarrow (10.7) \text{ has a solution.}$$

Note that formula (10.4) in Theorem 10.1 implies that $W_\omega(\tau)$ depends in a linear fractional way on τ.

(2)   Theorem 10.3 and numerical experiments say that

$$\rho_\omega(T) \triangleq \inf_{D \text{ comm. } \mathbf{Q}} \|DW_\omega(T(i\omega))D^{-1}\|$$

is an upper bound and a good approximation for $\mu_\mathbf{Q}(W_\omega(T(i\omega)))$.

Consequently, Doyle proposes that one replace (10.7) by

$$\sup_{\omega} \inf_{D(\omega)-\text{comm. } \mathbf{Q}} \|D(\omega)W_\omega(T(i\omega))D(\omega)^{-1}\| \leq 1.$$

One then attacks this by iteration. Guess T^K, to get the next guess T^{K+1}.

(1)   Minimize $\|D(\omega)W_\omega(T^k(i\omega))D(\omega)^{-1}\|$ over D's to get $D^0(\omega)$.

(2)   Since $D^0(\omega)W_\omega(\tau)[D^0(\omega)]^{-1}$ is linear fractional in τ, it has $\| \ \| \leq c$ if and only if τ is in a certain disk $\Delta_{Y(\omega)}$. Find a T in $\Delta_Y \cap \mathbf{A}$. Set $T_{K+1} = T$.

(3)   Go to (1).

Among the omitted details in the constraint of "internal stability." We won't go into what it means physically. Mathematically all that it does is impose interpolation constraints on T, so T is in $\mathbf{A}$ and meets a finite number of interpolation conditions. This causes no new problems in the solution.

---

[4]Actually $Q \triangleq C(1 + PC)^{-1}$ is better than T for parametrizing matrix-valued systems.

SOLUTION TO HOMEWORK EXERCISE 10.1. We want to show that

$$\{w \colon \|a + bw(1 - dw)^{-1}c\| \le 1\} = \Delta_Y$$

for some Y which we shall compute. For all $x \in \mathbb{C}^k$ we have

(10.8)
$$\|ax + bwy\|^2 \le \|x\|^2$$

where $y = (1 - dw)^{-1}cx$. Let p be the orthogonal projection onto the null space of c and choose a pseudoinverse m of c so that $mc + p = I$. Thus $m(1 - dw)y + pw = mcx + px = x$ and (10.8) becomes

$$\|ax + bwy\|^2 \le \|m(1 - dw)y + px\|^2,$$

$$\|[am(1 - dw) + bw]y + apx\|^2 \le \|m(1 - dw)y\|^2 + \|px\|^2 \quad \text{for all } y \text{ and } px.$$

$$([am(1 - dw) + bw]^*[am(1 - dw) + bw]y, y)$$
$$+ 2\,\mathrm{Re}([am(1 - dw) + bw]y, apx) + (a^*apx, px)$$
$$\le ([m(1 - dw)]^*[m(1 - dw)]y, y) + \|px\|^2.$$

The reader can carry on from here.

Part IV. The General H^∞ Optimization Problem

11. Nonlinear H^∞ Optimization

This final chapter treats the general (highly nonlinear) problem of optimizing over H^∞ or over $\mathbf{A}$. This problem was stated and motivated in Chapter 2 and a person who has read Chapter 3 and Chapter 2 only can make substantial headway in this section. To be somewhat self-contained this section contains some redundancy. Also in this section we shall only list results. One reason is that things are moving rapidly enough for methods to be constantly changing. Another more truthful reason is that I am tired of writing by now and anxious to finish. At any rate this chapter is fairly close to lectures given at Tel Aviv University in May of 1985. Those lectures were a part of a series called the Toeplitz Lectures which Tel Aviv University sponsors every two years. For more recent results see [**BHM**].

The key problem we treat here is:

Given a nonnegative-valued function $\Gamma(e^{i\theta}, z)$ of $e^{i\theta}$ and $z \in \mathbb{C}^N$ find

$$\text{(OPT)} \qquad\qquad \gamma^\theta = \inf_{f \in \mathbf{A}_N} \sup_\theta \Gamma(e^{i\theta}, f(e^{i\theta})).$$

For example when $N = 1$ if $\Gamma(e^{i\theta}, z) = |g(e^{i\theta}) - z|^2$ this is the original Nehari problem. If $z \in \mathbb{C}^4$ and

$$\Gamma(e^{i\theta}, z) = \left\| G - \begin{pmatrix} z_1 & z_2 \\ z_3 & z_4 \end{pmatrix} \right\|^2_{M_{2\times 2}},$$

then we get a matrix Nehari problem of the type solved in Part II. While Part II gave "explicit solutions" to (OPT) when its sublevel sets are "disks" in matrix space, the general (OPT) problem is too hard to solve explicitly. What we report on here is qualitative properties of the optimum when Γ is differentiable in $z \in \mathbb{C}^N$. The second example does not fit the mold precisely since matrix norms are not differentiable functions. Recall that $\mathbf{S}_\theta(c)$ denotes a sublevel set

$$\mathbf{S}_\theta(c) = \{z \in \mathbb{C}^N : \Gamma(e^{i\theta}, z) < c\}$$

of Γ.

What properties does a solution f_0 to OPT have? That is the subject of this talk. We start with z a single complex variable. Physically this means one designable parameter.

THEOREM 11.1 ([**H10**]). *Given $f_0 \in \mathbf{A}$ and $a(e^{i\theta}) \triangleq (\partial\Gamma/\partial z)(e^{i\theta}, f_0(e^{i\theta}))$ never equals zero, then f_0 is a strict local optimum if and only if*

$$\text{(i) } \Gamma(e^{i\theta}, f_0(e^{i\theta})) = \text{const.,} \qquad \text{(ii) } \text{wno}\, a > 0.$$

Here wno a stands for the winding number of a about the origin.

This is a strict generalization of the standard result for the classical case. The test would be trustworthy unless a dips close to the origin. That is, unless $\kappa \triangleq \inf_\theta |a(e^{i\theta})|$ is small. Then a small perturbation of a could change its winding number. This is not a major issue here, but in more complicated problems it matters.

Let's look at several designable parameters, $z \in \mathbb{C}^N$. If all functions involved are rational we get a nice generalization of Theorem I.

THEOREM 11.2 ([**H10**]). *Given $f_0 \in \mathbf{A}$ and*

$$a_j(e^{i\theta}) \triangleq (\partial\Gamma/\partial z_j)(e^{i\theta}, f_0(e^{i\theta}))$$

all rational and not simultaneously vanishing, then f_0 is a strict local optimum if and only if

(i) $\Gamma(e^{i\theta}, f_0(e^{i\theta})) = $ *constant;*

(ii) *Number of common zeros of a_j in disk—total number poles of a_j not counting overlaps in disk ≥ 0.*

EXAMPLE. $N = 1$. $^\#$zeros $a-^\#$poles $a = $ wno a. So this generalizes the $N = 1$ case exactly.

The previous theorems pertain to a function f_0 in $\mathbf{A}_N$ which we are testing for optimality. Suppose that f_0 is in H_N^∞, but we do not know that it is continuous. This indeed is the case which we have a priori in many situations, since an optimizing sequence f^k in $\mathbf{A}_N$ which is uniformly bounded (by normal families) has a cluster point f_0 in H_N^∞. What properties does an optimum f_0 in H^∞ have? The following is known to date.

Suppose $f_0 \in H_N^\infty$ is a solution to (OPT).

THEOREM 11.3 ([**HH**]). *If Γ varies continuously in θ, if the optimal sublevel sets $\mathbf{S}_\theta(\gamma_0) \subset \mathbb{C}^N$ of Γ are convex, nondegenerate, and uniformly bounded in θ, then $\Gamma(e^{i\theta}, f_0(e^{i\theta})) \equiv$ constant a.e.*

COROLLARY 11.4 ([**HH**]). *If Γ as above has strictly convex sublevel sets $\mathbf{S}_\theta$, then every local optimum f_0 is also a global optimum and is unique.*

PROOF OF COROLLARY. Suppose f_1 and f_2 are both local optima with $\|\Gamma(e^{i\theta}, f_1)\|_\infty = \|\Gamma(e^{i\theta}, f_2)\|_\infty = \gamma_0$; then $(f_1 + f_2)/2$ is also a local optimum (by convexity) and $\Gamma(e^{i\theta}, (f_1 + f_2)(e^{i\theta})/2) \equiv \gamma_0$ a.e. by Theorem 11.3; however, strict convexity implies that $(f_1 + f_2)(e^{i\theta})/2$ takes values strictly in the interior of $\mathbf{S}_\theta$ on a set θ of positive measure. Thus we have established uniqueness of a global optimum. That a local optimum f_1 must equal the global optimum follows by considering convex combinations $\lambda f_0 + (1 - \lambda)f_1$ and doing a simple (standard) estimate.

For $N = 1$, D. Schwartz, S. Warschawski, and I proved:

THEOREM 11.5 [**HSW**]. *When Γ is smooth with simply connected sublevel sets and $(\partial\Gamma/\partial z)(e^{i\theta}, f_0(e^{i\theta}))$ is "nonvanishing," then $\Gamma(e^{i\theta}, f_0(e^{i\theta})) = constant$ a.e.*

Subsequent to the Conference Board of the Mathematical Sciences conference, S. Hui (a student of Don Marshall) proved:

THEOREM 11.6 [**Hui**]. ($N = 1$). *Under the hypothesis of Theorem 11.3 plus the additional assumption that $\Gamma(e^{i\theta}, x + iy)$ is real analytic in x and continuous in y, it follows that the optimum f_0 in H^∞ is continuous.*

Now we return to the situation where f_0 is in $\mathbf{A}_N$. Trouble emerges when we look at the conditioning of Theorem 11.2. A small change in a_j's makes the number of common zeros of a_j equal to zero (in fact for generic a_j this equals zero). So the test generically fails. This is consistent with the original problem, since the generic f_0 is not an optimum.

So what we need in order to get a practical test which one could stably put on the computer is a "condition number" for a function f_0 to be optimum or at least a "condition number" for test (ii) in Theorem 11.2 to hold. Such a number κ will be zero if f_0 is optimum, be far away from zero if f_0 is far from optimum, etc. Some careful thought shows that the condition number appropriate for test (ii) is exactly

$$\kappa(a) = \sup_{\vec{h}} \left\{ \inf_\theta \left| \sum_j a_j h_j \right| : \vec{h} \in \overline{\mathbf{B}}\mathbf{A}_N \quad \text{wno} \sum_j a_j h_j \leq 0 \right\}.$$

Here $\mathbf{A}_N$ stands for the N-vector valued functions analytic and continuous on the disk, $\overline{\mathbf{B}}\mathbf{A}_N$ stands for the closed unit ball $\{\vec{h} \in \mathbf{A}_N : \sum_j |h_j|^2 \leq 1\}$, and $a = (a_1, \ldots, a_n)$. Complicated though it seems this exactly generalizes the κ for $N = 1$.

Now we must analyze it. Recall from Chapter 5.B the classical problem: given functions a_j in $\mathbf{A}$, find $h_j \in \mathbf{A}$ so that

(BI) $$\sum_j a_j h_j = 1.$$

Wiener showed that if the a_j's have no common zeros then a solution h_j exists, and Gelfand gave a very simple proof. A much more difficult problem is for $a_j, h_j \in H^\infty$. The hard part is that one actually must give an a priori bound C on the norm $\| \sum_{j=1}^N |h_j|^2 \|_\infty$ of the solutions h_j. This bound might be called the "Corona constant," and computing that it is finite is the meat of the classical Corona problem which L. Carleson solved. Recall that the Toeplitz Corona theorem in Chapter 5.B says:

TC THEOREM. $1/C = $ *smallest eigenvalue of* $\sum_j^N T_{a_j} T_{a_j}^*$.

Return to our original problem and let us do an

EXAMPLE. Suppose $a_j \in \mathbf{A}$. Then $\sum a_j h_j \triangleq q \in \mathbf{A}$, since $h_j \in \mathbf{A}$. We are assuming q is invertible, so wno $q \geq 0$. If we require wno $q \leq 0$, then $q \in \mathbf{A}$ and has wno $= 0$. Thus q and $1/q \in \mathbf{A}$. So $\sum a_j h_j / q = 1$ and h_j / q is a solution to (BI).

Now one can check through this construction to get $\kappa(a) = 1/C$. Thus computing the condition number $\kappa(a)$ is at least as hard as computing the Corona constant C. Another problem is that our functions a_j are not analytic, so we need a theorem which permits poles. Can we find such a generalization of the Toeplitz Corona theorem? This problem actually motivated the work in Chapter 5.B and prompted us to produce the more general TC Theorem (e.g., Theorem 5.B.1). Indeed what is called for here is more general than what can be found in Chapter 5.B and [**H10**] gives an explicit formula for $\kappa(a)$ which generalizes Theorem 5.B.1.

References

[**A**] W. B. Arveson, *Interpolation problems in nest-algebras*, J. Funct. Anal. **2** (1975), 208–233.

[**AAK1**] V. M. Adamjan, D. Z. Arov, and M. G. Kreĭn, *Analytic properties of Schmidt pairs for a Hankel operator and the generalized Schur-Takagi problem*, Math. USSR-Sb. **15** (1971), 31–73.

[**AAK2**] ____, *Infinite block Hankel matrices and their connection with the interpolation problem*, Akad. Nauk Armenia SSR Isvestia Mat. **6** (1971); English transl. in Amer. Math. Soc. Transl. (2) **111** (1978).

[**AAK3**] ____, *Infinite Hankel and generalized Carathéodory-Fejér and I. Schur problems*, Funktsional. Anal. i Prilozhen **2** (1968), 1–17. (Russian)

[**AAK4**] ____, *Bounded operators that commute with a contraction of class C_∞ of unit rank of nonunitary*, Funktsional. Anal. i Prilozhen **3** (1969), 86–87. (Russian)

[**AAK5**] ____, *Infinite Hankel block matrices and related continuation problems*, Izv. Akad. Nauk Armjan. SSR Ser. Mat. **6** (1971), 87–112. (Russian)

[**ACF1**] Gr. Arsene, Z. Ceauşescu, and C. Foiaş, *On intertwining dilations.* VII, Complex Analysis Joensuu 1978 (Proc. Colloq., Univ. Joensuu, Joensuu, 1978), Lecture Notes in Math., vol. 747, Springer-Verlag, 1979, pp. 24–45.

[**ACF2**] ____, *On intertwining dilations.* VIII, J. Operator Theory **4** (1980), 55–91.

[**A-D1**] D. Alpay and H. Dym, *Hilbert spaces of analytic functions, inverse scattering, and operator models.* I, Integral Equations and Operator Theory **7** (1984), 589–641.

[**A-D2**] ____, *On applications of reproducing kernel spaces to the Schur algorithm and rational J-unitary factorization*, Methods in Operator Theory and Signal Processing (editor I. Gohberg), Operator Theory: Advances and Applications **18** (1986).

[**AG**] Gr. Arsene and A. Gheondea, *Completing matrix contractions*, J. Operator Theory **7** (1982), 179–189.

[**An1**] T. Ando—has personally translated several of Potopov's basic papers on J-unitary functions to English.

[**An2**] T. Ando, *On a pair of commutative contractions*, Acta Sci. Math. (Szeged) **24** (1963), 88–90.

[**AnCF**] T. Ando, Z. Ceauşescu, and C. Foiaş, *On intertwining dilations*. II, Acta Sci. Math. (Szeged) **39** (1977), 3–14.

[**Ar**] D. Z. Arov, *Darlington's method for dissipative systems*, Dokl. Akad. Nauk SSSR **201** (1971), 559–562; English transl. in Soviet Physics Dokl. **16** (1971).

[**Av**] Y. Avniel, *Realization and approximation of stationary stochastic processes*, Thesis, Mass. Inst. of Tech., Cambridge, Mass., 1984.

[**AY**] A. C. Allison and N. J. Young, *Numerical algorithms for Nevanlinna-Pick problem*, Numer. Math. **42** (1983), 125–145.

[**B1**] J. A. Ball, *Interpolation problems of Pick-Nevanlinna and Löewner types for meromorphic matrix functions*, Integral Equations Operator Theory **9** (1986), 155–203.

[**B2**] _____, *Models for non-contractions*, J. Math. Anal. and Appl. **52** (1975), 235–254.

[**B3**] _____, *Nevanlinna-Pick interpolation: Generalizations and applications*, Proc.Asymmetric Algebras and Invariant Subspaces Conference, Indiana Univ. (March 1986) (to appear).

[**BC**] J. Brenner and L. Cummings, *The Hadamard maximum determinant problem*, Amer. Math. Monthly **79** (1972), 626–630.

[**Bog**] J. Bognar, *Indefinite inner product spaces*, Springer-Verlag, 1974.

[**BG1**] J. A. Ball and I. Gohberg, *A commutant lifting theorem for triangular matrices with diverse applications*, Integral Equations Operator Theory **8** (1985), 205–267.

[**BG2**] _____, *Shift invariant subspaces, factorization, and interpolation for matrices*. I: *The canonical case*, Linear Algebra Appl. **74** (1986), 87–150.

[**BG3**] _____, *Classification of shift invariant subspaces and noncanonical factorizations*, Linear and Multilinear Algebra **20** (1986), 27–61.

[**BG4**] _____, *Classification of shift invariant subspaces with Hermitian form and completion of matrices*, Proc. Workshop on Operator Theory and Applications (Amsterdam 1985), I. Gohberg and M. A. Kaashock, editors, Operator Theory: Adv. Appl., vol. 19.

[**BGK1**] H. Bart, I. Gohberg, and M. A. Kaashoek, *Minimal factorization of matrix and operator functions*, Birkhäuser, 1979.

[**BGK2**] _____, *Wiener-Hopf factorization of analytic operator functions and realization* (to appear).

[**BH1**] J. A. Ball and J. W. Helton, *A Beurling-Lax theorem for the Lie group* U(m, n) *which contains most classical interpolation theory*, J. Operator Theory **9** (1983), 107–142.

[**BH2**] _____, *Beurling-Lax representations using classical Lie groups with many applications*. II: GL($n, \mathbb{C}$) *and Wiener-Hopf factorization*, Integral Equations Operator Theory **7** (1984), 291–309.

[**BH3**] ____, *Beurling-Lax representations using classical Lie groups with many applications.* III: *Groups preserving two bilinear forms*, Amer. J. Math. **108** (1986), 95–174.

[**BH4**] ____, *Beurling-Lax representations using classical Lie groups with many applications.* IV: $GL(n, R), SL(n, \mathbb{C})$, *and a solvable group*, J. Funct. Anal. **69** (1986), 178–206.

[**BH5**] ____, *Interpolation problems of Pick-Nevanlinna and Löewner types for meromorphic matrix functions*: *Parametrization of the set of all solutions*, Integral Equations Operator Theory **9** (1985), 155–203.

[**BH6**] ____, *Interpolation with outer functions and gain equalization in amplifiers*, Proceedings of Mathematical Theory of Methods and Systems, (MTNS Conference, Delft, 1979).

[**BH7**] ____, *Lie groups over the field of rational functions, signed spectral factorization, signed interpolation, and amplifier design*, J. Operator Theory **8** (1982), 19–64.

[**BH8**] ____, *Factorization results related to shifts in an indefinite metric*, Integral Equations Operator Theory **5** (1982), 632–658.

[**BHM**] J. Bence, J. W. Helton, and D. Marshall, H^∞ *optimization*, IEEE Conference on Decision and Control, Athens (1986) (to appear).

[**BJ1**] W. Barrett and C. R. Johnson, *Determinantal formulae for matrices with sparse inverses*, Linear Algebra and Its Appl. **56** (1984), 73–88.

[**BJ2**] ____, *Spanning-tree extensions of the Hadamard-Fischer inequalities*, Linear Algebra and Its Appl. **66** (1985), 177–193.

[**Bo**] J. Border, *Nonlinear Hardy spaces and electrical power transfer*, Thesis, University of California, San Diego, Calif., 1979.

[**BP**] A. Bloch and G. Polya, *Abschätzung des Beitrages einer Determinante*, Vjschr. naturfosch. Ges. Zür **78** (1933), 27–33.

[**BR1**] J. A. Ball and A. C. M. Ran, *Hankel norm approximation for rational matrix functions in terms of realizations*, Math. Theory of Networks and Systems Proc. (Stockholm, 1985), Modelling, Identification and Robust Control, C. I. Brynes and A. Lindquist, editors, North-Holland, 1986, pp. 285–296.

[**BR2**] ____, *Optimal Hankel norm model reductions and Wiener-Hopf factorization*. I: *The canonical case*, SIAM J. Control and Optim. **25** (1987), 362–382.

[**BR3**] ____, *Global inverse spectral problems for rational matrix functions*, Linear and Multilinear Algebra **86** (1987), 237–282.

[**BR4**] ____, *Local inverse spectral problems for rational matrix functions*, Integral Equations Operator Theory **10** (1987) (to appear).

[**BSS**] M. Bettayeb, L. Silverman, M. G. Safonov, Proc. IEEE Conference on Decision and Control (1980).

[**C1**] D. Clark, *On the spectra of bounded Hermitian, Hankel matrices*, Amer. J. Math. **90** (1968), 627–656.

[**C**] W. Chen, *Theory and design of broadband matching networks*, Pergamon Press, 1976.

[**CF1**] Z. Ceauşescu and C. Foiaş, *On intertwining dilations*. V, Acta Sci. Math. (Szeged) **40** (1978), 9–32.

[**CF2**] ____, *On intertwining dilations*. VI, Rev. Roumaine Math. Pures Appl. **23** (1978), 1471–1482.

[**CFS**] S. Campbell, G. Faulkner, and R. Sine, *Isometries, projections and Wold decompositions in operator theory and functional analysis* (I. Erdelyi, ed.), Res. Notes in Math., vol. 38, Pitman, 1979, pp. 85–114.

[**CG**] K. Clancy and I. Gohberg, *Factorization of matrix functions and singular integral operators*, Birkhäuser, 1982.

[**D**] R. G. Douglas, *Banach algebra techniques in operator theory*, Academic Press, 1972.

[**Da**] R. D. Daniels, *Approximation methods for electronic filter design*, McGraw-Hill, New York, 1974.

[**DD**] P. DeWilde and H. Dym, *Lossless: Inverse scattering, digital filters, and estimation theory*, IEEE Trans. Inform. Theory **IT-30** (1984), 644–662.

[**DGo**] H. Dym and I. Gohberg, *Extensions of band matrices with band inverses*, Lin. Alg. and Its Applications **36** (1981), 1–24.

[**deBr1**] L. de Branges, *Some Hilbert space of analytic functions*. I, Trans. Amer. Math. Soc. **106** (1963), 445–468.

[**deBr2**] ____, *Some Hilbert spaces of analytic functions*. II, J. Math. Anal. Appl. **11** (1965), 44–72.

[**deBr3**] ____, *Some Hilbert spaces of analytic functions*. III, J. Math. Anal. Appl. **12** (1965), 149–186.

[**deBr4**] ____, *Complementation in Krein space*, Preprint.

[**deBr5**] ____, *Krein spaces of analytic functions*, Preprint.

[**Do**] W. F. Donoghue, *Monotone matrix functions*, Springer-Verlag, 1974.

[**Dor**] R. C. Dorf, *Modern control systems*, Addison-Wesley, 1981.

[**Doy1**] J. C. Doyle, *Analysis of feedback systems with structured uncertainties*, Proc. IEEE-D **129** (1982), 242–250.

[**Doy2**] ____, *Synthesis of robust controllers*, Proc. CDC (1983), 109–114.

[**DG**] C. A. Desoer and C. L. Gustafson, *Design of multivariable feedback system with simple unstable plant*, Berkeley ERL Memorandum No. M82/60.

[**DGK**] P. H. Delasarte, Y. Genir, and Y. Kemp, *The Nevanlinna-Pick problem for matrix-valued functions*, SIAM J. Appl. Math. **36** (1979), 47–61.

[**DH**] R. G. Douglas and J. W. Helton, *Inner dilations of analytic matrix functions and Darlington synthesis*, Acta Sci. Math. (Szeged) **34** (1973), 61–67.

[**DKW**] C. Davis, W. M. Kahan, and H. F. Weinberger, *Norm-preserving dilations and their applications to optimal error bounds*, SIAM J. Numer. Anal. **19** (1982), 445–469.

[**DLMS**] C. Desoer, R. W. Liu, J. Murray, and R. Saeks, *Feedback system design: The fractional representation approach to analysis and synthesis*, IEEE Trans. Automat. Control **AC-25** (1980), 399–412.

[DMP] R. G. Douglas, P. S. Muhly, and C. Pearcy, *Lifting commuting operators*, Michigan Math. J. **15** (1968), 385–395.

[DS] J. C. Doyle and G. Stein, *Multivariable feedback design: Concepts for a classical modern synthesis*, IEEE Trans. Automat. Control **AC-26** (1981), 4–16.

[DSa] J. C. Doyle and M. S. Safanov, *Convexity of the block diagonal sealing problem*, Internal Memo, Honeywell Systems and Research Center, Minneapolis, Minn., 1982.

[DSW] J. C. Doyle, G. Stein, and J. Wall, *Performance and robustness for structured uncertainty*, IEEE Conference on Decision and Control (1982).

[Du1] P. L. Duren, *Theory of* H^P *spaces*, Academic Press, New York, 1970.

[Du2] _____ , *Univalent functions*, Springer-Verlag, New York, 1983.

[DVK] P. DeWilde, A. Vieira, and T. Kailath, *On a generalized Szegö-Levinson realization algorithm for optimal linear predictors based on a network synthesis approach*, Special Issue on Math. Foundations of Systems Theory, IEEE Trans. Circuit Theory **25** (1978), 663–675.

[Dym] H. Dym, Lecture notes from an NSF regional conference in 1984 on J-unitary matrix functions, in preparation.

[EP] Efimov and Potapov, J-*expanding functions and their role in the analytic theory of electrical circuits*, Uspekhi Mat. Nauk **28** (1973), 65–130; English transl. in Russian Math. Surveys **281** (1973).

[F] A. E. Frazho, *Three inverse scattering algorithms for the lifting theorem*, Preprint.

[FD] B. A. Francis and J. Doyle

[Fd1] I. I. Fedchin, *Description of solutions of the tangential Nevanlinna-Pick problem*, Akad. Nauk Armjan SSR Dokl. **60** (1975), 37–42. (Russian)

[Fd2] _____ , *Tangential Nevanlinna-Pick problem with multiple points*, Akad. Nauk Armjan SSR Dokl. **64** (1975), 214–218. (Russian)

[Fe] A. Feintuch, Discrete-time feedback systems: An operator theoretic approach.

[FeTa] A. Feintuch and A. Tannenbaum, *On the sensitivity minimization problem for linear time-varying periodic systems*, SIAM J. Control Optim. **24** (1986), 1076–1085.

[FF1] A. Feintuch and B. Francis, *Uniformly optimal control for time varying systems*, Systems Control Lett. (1984).

[FF2] _____ , *Uniformly optimal control for linear systems*, Automatica **21** (1985), 563–574.

[FF3] _____ , *Minimum sensitivity problems for time-varying systems*, Proc. M.T.N.S., 1985.

[FF4] _____ , *The general distance problem for linear periodic systems*, Preprint.

[FFr1] C. Foiaş and A. E. Frazho, *Redheffer products and lifting contractions on Hilbert space*, J. Operator Theory **11** (1984), 193–196. (Other articles on interpolation and H^∞ approximation by the same authors will appear.)

[FFr2] ——, *On the Schur representation in the commutant lifting problem.*
I, Integral Equations and Operator Theory.

[FHZ] B. A. Francis, J. W. Helton, and G. Zames, *Optimal feedback controllers for linear multivariable systems*, IEEE Trans. Automat. Control **29** (1984), 888–900.

[Foi1] C. Foiaş, *Some applications of structural models for operators on Hilbert spaces*, Proc. Intern. Congr. Math. (Nice, September 1970), Tome **2**, Gauthier-Villars, Paris, (1971), 433–440.

[Foi2] ——, *Contractive intertwining dilations and waves in layered media*, Proc. Internat. Congr. Math. (Helsinki, 1978), Vol. 2, Acad. Sci. Fennica, Helsinki, 1980, pp. 605–613.

[Foi3] ——, Lecture at the International Congress of Mathematics, Helsinki, 1978.

[FO] B. A. Francis and S. O'Young, Preprint.

[FPE] G. F. Franklin, J. D. Powell, and A. Emami-Naeini, *A first course in feedback control*, Stanford University, 1984.

[Fr] B. A. Francis, *A course in H^∞ control theory*, Lecture Notes in Control and Information Sci., vol. 88, Springer-Verlag, 1986.

[FS] Y. Foures-Bruhat and I. E. Segal, *Causality and analyticity*, Trans. Amer. Math. Soc. **78** (1955), 385–405.

[FSk] A. Feintuch and R. Saeks, *System theory: A Hilbert space approach*, Academic Press, New York, 1982.

[FeiT] ·A. Feintuch and A. Tannenbaum, *Gain optimization for distributed plants*, Systems Control Lett. **6** (1986), 295–301.

[FT] M. K. H. Fan and A. L. Tits, *Characterization and efficient computation of the structured singular value*, Preprint.

[Fu] P. A. Fuhrmann, *Linear systems and operators in Hilbert space*, McGraw-Hill, New York, 1981.

[FV] B. A. Francis and M. Vidyasagar, *Algebraic and topological aspects of the regulator problem for lumped linear systems*, Automatica **13** (1983), 87–90.

[G] J. Garnett, *Bounded analytic functions*, Academic Press, New York, 1981.

[GB] I. Gohberg and M. A. Barker, *On factorization of operators relative to a discrete chain of projections*, Amer. Math. Soc. Transl. (2) **90** (1970), 81–103.

[GD] C. L. Gustafson and C. A. Desoer, *Controller design for linear multivariable feedback systems with stable plants, using optimization with inequality constraints*, Internat. J. Control **37** (1983), 881–907.

[GJSW1] R. Grone, C. R. Johnson, E. de Sá, and H. Wolkowicz, *Positive definite completions of partial Hermitian matrices*, Linear Algebra Appl. **58** (1984), 109–124.

[GJSW2] ——, *Improving Hadamard's inequality*, Linear and Multilinear Algebra **16** (1984), 305–322.

[Gl] K. Glover, *All optimal Hankel-norm approximations of linear multivariable systems and their L^∞-error bounds*, Internat. J. Control **39** (1984), 1115–1193.

[GLR] I. Gohberg, P. Lancaster, and L. Rodman, *Matrices and indefinite scalar products*, Birkhäuser and Springer-Verlag, 1983.

[Go] M. Golumbic, *Algorithmic graph theory and perfect graphs*, Academic Press, 1980.

[GW1] R. Goodman and N. R. Wallach, *Classical and quantum mechanical systems of Toda-lattice type*, Comm. Math. Phys. **94** (1984), 177–217.

[GW2] R. Goodman and N. R. Wallach, *Structure and unitary cocycle representations of loop groups and the group of diffeomorphisms of the circle*, J. Reine Angew. Math. **346** (1984), 69–133.

[H1] J. W. Helton, *Operator theory and broadband matching*, Proc. Allerton Conf. Circuits and Systems Theory, 1976, pp. 91–98.

[H2] ____, *A simple test to determine gain bandwidth limitations*, Proc. IEEE Internat. Conf. on Circuits and Systems (Phoenix, 1977), IEEE, New York, 1977, pp. 628–631.

[H3] ____, *Orbit structure of the Möbius transformation semigroup action on H^∞ broadband matching*, Adv. in Math. Suppl. Stud., vol. 3, Academic Press, New York, 1978, pp. 129–197.

[H4] ____, *The distance of a function to H^∞ in the Poincaré metric; electrical power transfer*, J. Funct. Anal. **38** (1980), 273–314.

[H5] ____, *Broadbanding: Gain equalization directly from data*, IEEE Trans. Circuits and Systems **28** (1981), 1125–1137.

[H6] ____, *Non-Euclidean functional analysis and electronics*, Bull. Amer. Math. Soc. (N.S.) **7** (1982), 1–64.

[H7] ____, *A systematic theory of worst case optimization in the frequency domain; high frequency amplifiers*, IEEE Internat. Conf. on Circuits and Systems (Newport Beach, 1983), IEEE, New York, 1983, 1052–1054.

[H8] ____, *An H^∞ approach to control*, IEEE Conf. on Decision and Control, 1983 (San Antonio).

[H9] ____, *Power spectrum reduction by optimal Hankel norm approximation of the phase of the outer spectral factor*, IEEE Trans. Automat. Control **30** (1985), 1192–1201.

[H10] ____, *Optimization over H^∞ and the Toeplitz Corona theorem*, J. Operator Theory **15** (1986), 359–375.

[Hi] W. Hintzman, *Best uniform approximations via annihilating measures*, Bull. Amer. Math. Soc. **76** (1970), 1062–1066.

[Hls] H. Helson, *Invariant subspaces*, Academic Press, New York, 1964.

[HH] J. W. Helton and R. Howe, *A bang-bang principle for the frequency domain*, J. Approx. Theory **47** (1986), 101–121.

[HJ] R. Horn and C. R. Johnson, *Matrix analysis*, Cambridge Univ. Press, New York, 1985.

[HP] Hewlett-Packard, *S-parameter design*, Application Note 154, 1972.

[HR] J. H. C. Van Heuven and T. Z. Rossi, *The invariant properties of a multivalue n-port in linear embedding*, IEEE Trans. Circuits and Systems **19** (1972), 176–183.

[HSp] J. W. Helton and R. Speciale, *A complete and unambiguous solution to the super-TSD multiport-calibration problem*, IEEE Conf. on Microwave Circuits (Boston).

[Hui] S. Hui, Thesis, University of Washington, Seattle, Wash., 1986.

[Iv] T. S. Ivanchenko, *The Schur problem in the case of indefinite metrics*, Dokl. Akad. Nauk Ukrain. SSR Ser. A **5** (1980), 8–13. (Russian)

[J] C. R. Johnson, *The many proofs of Hadamard's determinantal inequality*, manuscript.

[JH] E. Jonckheere and J. W. Helton, IEEE Trans. Automat. Control **30** (1985), 1192–1201.

[JM] C. R. Johnson and T. Markham, *Compression and Hadamard power inequalities for M-matrices*, Linear and Multilinear Algebra **18** (1985), 23–34.

[JMP] C. R. Johnson, R. Merris, and S. Pierce, *Inequalities involving immanants and diagonal products for H-matrices and positive definite matrices*, Portugal. Math. vol. 43, Fasc. 1, 1985/86.

[JN] C. R. Johnson and M. Newman, *A surprising determinantal inequality for real matrices*, Math. Ann. **247** (1980), 176–186.

[JNP] H. M. James, N. B. Nichols, and R. S. Phillips, *Theory of servomechanisms*, M.I.T. Rad. Lab. Series, vol. 25, McGraw-Hill, New York, 1974.

[JR1] C. R. Johnson and L. Rodman, *Inertia possibilities for completions of partial Hermitian matrices*, Linear and Multilinear Algebra **16** (1984), 179–195.

[JR2] ——, *Completion of partial matrices to contractions*, J. Funct. Anal. **69** (1986).

[K] S. Y. Kung, IEEE Trans. Circuits and Systems.

[Ka] T. Kailath, *Linear systems*, Prentice-Hall, Englewood Cliffs, N. J., 1980.

[KL] J. Kraus and D. Larson, *Reflexivity and distance formulae*, Proc. London Math. Soc. (to appear).

[Ko] P. Koosis, *Introduction to H^p spaces*, London Math. Soc. Lecture Note Ser. **40**, Cambridge Univ. Press, New York, 1980.

[KP1] P. P. Khargonekar and K. R. Poolla, *Robust stabilization of distributed systems*, Automatica—J. IFAC (1986).

[KP2] ——, *Uniformly optimal control of linear time-invariant plants: Nonlinear time-varying controllers*, submitted for publication.

[KPT] P. P. Khargonekar, K. R. Poolla, and A. Tannenbaum, *Robust control of linear time invariant plants using periodic compensation*.

[KPot] I. V. Kovalishyn and V. P. Potapov, *Indefinite metric in the Pick-Nevanlinna problem*, Akad. Nauk Armyan SSR Dokl. **59** (1974), 3–9. (Russian)

[KR] E. Kuh and Roher, *Theory of linear active networks*, Holden-Day, San Francisco, Calif., 1967.

[**KT**] P. Khargonekar and A. Tannenbaum, *Non-Euclidean metrics and the robust stabilization of systems with parameter uncertainty*, IEEE Trans. Automat. Control **30** (1985), 8575–8578.

[**Kw**] Kwakernaak, *Robustness optimization of linear feedback systems*, IEEE Conf. on Decision and Control, December, 1983.

[**M**] D. Marshall, *An elementary proof of the Pick-Nevanlinna interpolation theorem*, Michigan Math. J. **21** (1974).

[**Nu1**] A. A. Nudelman, *On a new problem of the type of moment problem*, Dokl. Akad. Nauk SSSR **233** (1977), 792–795.

[**Nu2**] ____, *On a generalization of classical interpolation problems*, Dokl. Akad. Nauk SSSR **256** (1981), 790–793. (Russian)

[**Nu3**] ____, *On a generalization of classical interpolation problems*, Dokl. Akad. Nauk SSSR **256** (1981), 790–793.

[**NF**] B. Sz.-Nagy and C. Foiaş, *Harmonic analysis of operators on Hilbert space*, North-Holland, Amsterdam, 1970.

[**NF1**] ____, *On contractions similar to isometries and Toeplitz operators*, Ann. Acad. Sci. Fenn. Ser. A I Math. **2** (1976), 553–564.

[**NS1**] A. M. Nīkolaĭčuk and I. M. Spītkovs′kiĭ, *The Riemann boundary value problem with a Hermitian matrix*, Dokl. Akad. Nauk SSSR **221** (1975), 1280–1283; English transl. in Soviet Math. Dokl. **16** (1975), 533–536.

[**NS2**] ____, *Factorization of Hermitian matrix functions and their applications to boundary value problems*, Ukrainian Math. J. **27** (1975), 767–779.

[**O**] K. Ogato, *Modern control engineering*, Prentice-Hall, Englewood Cliffs, N. J., 1970.

[**P**] Potapov. Ando has translated several of his papers.

[**Pr**] S. C. Powers, *Operators and function theory*, Reidel, Dordrecht, Hingham, Mass., 1985.

[**PW**] E. Polack and Y. Wardi, *Nondifferentiable optimization algorithm for designing control systems having singular value inequalities*, Automatica—J. IFAC **18** (1982), 267–283.

[**RR1**] M. Rosenblum and J. Rovnyak, *The factorization problem for nonnegative operator valued functions*, Bull. Amer. Math. Soc. (N.S.) **77** (1981), 287–318.

[**RR2**] ____, *Hardy classes and operator theory*, Oxford University Press, 1986.

[**S1**] D. Sarason, *Generalized interpolation in H^∞*, Amer. Math. Soc. Transl. **127**, Amer. Math. Soc., Providence, R. I., 1967, pp. 180–203.

[**S2**] ____, *Operator theoretic aspects of the Nevanlinna-Pick interpolation problem in operators and function theory*, Lancaster Conference Notes, Reidel, Dordrecht, Hingham, Mass., 1985, pp. 279–314.

[**Sc**] A. Scales, Interpolation with meromorphic functions of minimal norm, Ph.D. Dissertation, University of California, San Diego, La Jolla, Calif., 1982.

[**SS**] R.Saucedo and E. E. Schering, *Introduction of continuous and digital control systems*, Macmillan, 1968.

[**SW**1] G. Segal and G. Wilson, *Loop groups and equations of* KdV *type*, Inst. Hautes Études Sci. Publ. Math. **61** (1985), 5–65.

[**SW**2] ____, *Loop groups and their representations*, forthcoming monograph.

[**SZ**1] B. Schwarz and A. Zaks, *Matrix Möbius transformations*, Comm. Algebra **9** (1981), 1913–1968.

[**SZ**2] ____, *Higher dimensional Euclidean and hyperbolic matrix spaces*, J. Analyse Math. **46** (1986), 271–282.

[**T**1] A. Tannenbaum, *Modified Nevanlinna-Pick interpolation and plants with uncertainty in the gain factor*, Internat. J. Control (1982).

[**T**2] ____, *On the multivariable gain margin problem*, Automatica—J. IFAC (to appear).

[**T**3] ____, *Invariance and system theory: Algebraic and geometric aspects*, Lecture Notes in Math., vol. 845, Springer-Verlag, 1981.

[**Vek**] N. P. Vekua, *Systems of integral equations*, Noordhoff, 1953.

[**WS**1] H. Wolkowicz and G. Styan, *Bounds for eigenvalues using traces*, Linear Algebra Appl. **29** (1980), 471–506.

[**WS**2] ____, *More bounds for eigenvalues using traces*, Linear Algebra Appl. **31** (1980), 1–17.

[**Y**] N. Young, *Interpolation by analytic matrix functions, operator and function theory*, Lancaster Conference Notes, Reidel, Dordrecht, Hingham, Mass., 1985, pp. 351–383.

[**Yeh**1] Fang-Bo Yeh, *Model reduction problem in theory and design in multivariable systems*, Lecture Notes, Cheng-Kung University Aeronautics and Astronautics Institute, Taiwan, 1984–1985.

[**Yeh**2] ____, *Numerical solution of matrix interpolation problem*, Ph.D. Thesis, University of Glasgow, Glasgow, Scotland, 1983.

[**Yeh**3] ____, *One step extension method of matrix Nevanlinna-Pick problem*, in preparation, 1985.

[**Yeh**4] ____, *Singular value decomposition of infinite block Hankel matrix*, in preparation, 1985.

[**YJB**] D. C. Youla, H. A. Jabr, and J. J. Bongiorno, *Modern Wiener-Hopf design of optimal controllers*. II, IEEE Trans. Automat. Control **21** (1976), 319–338.

[**YS**] D. C. Youla and M. Saito, *Interpolation with positive real functions*, J. Franklin Inst. **284** (1967), 77–108.

[**Z**] G. Zames, *Feedback and optimal sensitivity: Model reference transformations, multiplicative seminorms, and approximate inverses*, IEEE Trans. Automat. Control **26** (1981), 301–320.

[**ZF**] G. Zames and B. A. Francis, *Feedback and minimax sensitivity*, Advanced Group for Aerospace Research and Development NATO Lectures Notes, No. 117, Multivariable Analysis and Design Techniques.